LOCAL-AREA NETWORKS WITH FIBER-OPTIC APPLICATIONS

LOCAL-AREA NETWORKS WITH FIBER-OPTIC APPLICATIONS

Donald G. Baker

A RESTON BOOK
Prentice-Hall
Englewood Cliffs, New Jersey

Library of Congress Cataloging-in-Publication Data

Baker, Donald G.
  Local-area networks with fiber-optic applications.

  "A Reston book."
  Includes bibliographies and index.
  1. Local area networks (Computer networks) 2. Fiber
optics. I. Title.
TK5105.7.B35 1986   004.6'5   85-31537
ISBN 0-8359-4100-0

A Reston Book
Published by Prentice-Hall
A division of Simon & Schuster, Inc.
Englewood Cliffs, New Jersey 07632

© 1986 by
Prentice-Hall
Englewood Cliffs, New Jersey 07632

*All rights reserved. No part of this book may be
reproduced in any way, or by any means, without
permission in writing from the publisher.*

10 9 8 7 6 5 4 3 2 1

PRINTED IN THE UNITED STATES OF AMERICA

To my wife, Barbara,
and to my daughters, Patricia and Stephanie,
for their patience and understanding

CONTENTS

Acknowledgments   xi

Preface   xiii

Chapter 1
INTRODUCTION   1
        Overview   2

Chapter 2
INTERNATIONAL STANDARDS ORGANIZATION NETWORK MODEL   9
        Physical Layer   13
        Data-Link Layer   13
        Network Layer   14
        Transport Layer   14
        Session Layer   15
        Presentation Layer   15
        Application Layer   16

Chapter 3
COMMON NETWORKS   19
        Packet Switching   20
        ARPANET   21

IBM's Systems Network Architecture: SNA 22
Digital Equipment Corporation Network: DECNET 24
Xerox Corporation: Ethernet 24
Wangnet 26
The X.25 29
Review Problems 30

Chapter 4
NETWORK TOPOLOGY 31

Introduction 33
Graph Theory 37
Network Flow 38
Time-Delay Analysis 48
Channel Capacity 55
Noise Considerations 56
Backbone Design 62
Review Problems 63
References 64

Chapter 5
PHYSICAL LAYER 65

Introduction to LAN Information Theory 67
Cable Plant Design 78
Ethernet Physical Layer Characteristics 84
Functional Description 90
Broadband Cable Plants 92
Fiber-Optic Cable Plants 97
Fiber-Optic Transmitters and Receivers 106
Comparisons Between Fiber Optics and Copper Cable
    Technology 112
Cyclic Redundancy Checking 114
Receiver Decoders 121
Multiplexers 122
Review Problems 126
References 128

Chapter 6
DATA-LINK LAYER 129

Elementary Protocol 131
High-Level Data-Link Control (HDLC) and Synchronous Data-Link Control
    (SDLC) 148

Contents ix

          ARPANET   150
          SNA and X.25   152
          DECNET   152
          Ethernet   153
          Protocol Specifications and Verification   155
          References   158

Chapter 7
NETWORK LAYER   159

          Virtual Circuits   160
          Datagrams   161
          Routing Techniques   163
          Local-Area Networks   171
          Collision-Sensing Networks   172
          Ring CSMA   172
          CSMA/CD Bus Network: Ethernet   182
          DELNI: Local Network Interconnect   184
          Transceivers   184
          Servers   184
          Token-Passing Networks   185
          Star Networks   191
          Hybrid Star Network   197
          Gateways   201
          Broadband Cable Networks   205
          Broadband Fiber-Optic Networks   207
          Overview of Packet Radio and Satellite   209
          Selection of Local Area Networks   211
          Review Problems   212
          References   212

Chapter 8
TRANSPORT LAYER   213

          Transport Error Recovery   217
          Network Security   220
          Transport-Layer Efficiency   221
          Transport-Layer Hardware Issues   222
          Virtual Terminals   222
          Network Servers   223
          ARPANET   223
          DECNET   228
          CCITT X.25 Packet-Switching Network   228
          References   228

Chapter 9
SESSION LAYER   229
        SNA Session   231
        Session Discussion   234
        SYTEK Session Layer   235
        References   236

Chapter 10
PRESENTATION LAYER   237
        File Transfer Protocols   238
        Virtual Terminals   240
        Security Without Encryption   246
        Security with Encryption   247
        Attacks on Encryption   248
        Data and Text Compression   251
        Conclusion   253
        References   254

Chapter 11
APPLICATIONS LAYER   255
        Distribution Data Base   257
        Homogeneous Operating Systems   270
        Parallel Processing   272
        References   279

Index   281

ACKNOWLEDGMENTS

The following people contributed to the completion of this book: Maria Vigneau, who typed most of the manuscript on her off hours and helped immeasurably with her assistance and skill; and Mrs. Carol Wood, who was responsible for editing the manuscript prior to delivery to the publisher. I wish to express my gratitude and thanks.

PREFACE

Local-area networks (LANs) are a technique of interconnecting equipment such as printers, terminals, computers, facsimile, and other devices in a network. These devices are accessible to multiple users, which allows resource sharing and thus efficient use of all equipment on the LAN. Most present-day network implements have microprocessors to process data prior to transmission due to their low cost.

The reader will gain a working knowledge of LANs and their applications from this text. These applications will be presented both in copper and fiber-optics technologies and the various trade-offs explored. Some of the more common integrated circuits used for networks are presented, and circuit diagrams included.

This book is a useful reference for engineers who wish to become acquainted with LANs in a rather painless manner. The LANs designed with fiber optics will introduce the reader to the technology without heavy involvement in physics or component design, such as for transmitters, receivers, multiplexers, and star couplers. Many books can be used to augment the text presentation to give the reader more information on the subject. Chapter references provide lists for further reading.

This book can be used as an introductry text on local area networks for five-day seminars or as a senior-level college course text.

The reader should have some background in electronics, particularly in digital circuits. To extend one's knowledge in fiber optics and understand other books on LANs, the reader must have a good background in calculus, but this text is not that stringent.

The book is arranged in two levels of complexity. Chapters 1 and 2 give the reader a brief overview of LAN technology and will answer the following questions: What are classified as local-area networks? Why are LANs useful? What is their future outlook? These chapters explain the International Standards Organization (ISO) model, which is widely accepted by industry. Chapter 3 is devoted to exploring some of the common LANs and applying the ISO model where possible.

Chapter 4 gives the reader some of the tools necessary for both the analysis and design of LANs networks. The study will contain design information on several of the more common network topologies, such as the ring and bus star, and other miscellaneous topics.

Chapters 5 through 11 cover the design and analysis of LANs; this is again a discussion of the ISO model in finer detail with more calculations.

Chapter 5, on the physical layer, is hardware oriented; this chapter delves into actual physically connected networks with a rich assortment of circuit diagrams. Both copper and fiber-optic cable plant designs are investigated, and performance and cost trade-offs are examined. The chapter is rounded out with several design examples.

Chapter 6, on the data-link layer, covers various common protocols that are widely accepted by industry. Analysis and design are investigated along with test and protocol testing. Some of the more common controllers are also addressed here.

The next topic in the ISO model hierarchy is the network layer, Chapter 7. The discussion here examines various LAN techniques and the analysis of some of the popular networks in use today. This chapter concludes the hardware issues, and the remainder of the book is devoted to software.

The transport and session layers (Chapters 8 and 9) are covered mainly through examples of existing LANs. Some design material is also given in Chapter 8.

The discussion of topics in the presentation layer (Chapter 10) is structured to cover issues of current importance. Terminals, due to the microprocessor impact, have a large number of added features that are examined.

The application layer (Chapter 11) is not covered in any detail because it is dependent on a particular user, but some state-of-the-art examples are given.

This book has an example of a LAN design from start to finish, with a part embedded in each ISO layer, chapter by chapter. The design consists of a token-passing fiber-optic LAN functioning as a distributed computer. The ring has implements such as a computer, disk storage station, terminals, and gateway station. A very detailed design of a telephone and foreign gateway station is also presented.

Often overlooked in other texts is the reliability issue. In many military and critical industrial applications, this topic is of prime importance. In military procurement for certain noncritical combat equipment, it is becoming more common to seek ruggedized industrial-grade equipment. Often military-grade equipment is obsolete shortly after it is released for production because of the rapid changes in technology.

A topic for future consideration, parallel processor controllers, is discussed in various chapters. The investigation is based on the Intel iAPX 432 and bit-slice technologies. A design example is examined in detail.

Some board-level components are discussed to give the reader an appreciation of what is available in the marketplace. Computers for controllers, for example, need not be designed due to the large variety of single-board computers available unless there is an equipment space problem or run time is too slow. Bus systems such as the S100, Q bus, and multibus are widely accepted by industry; these are also examined.

The design example presented uses a controller that is designed both with a microprocessor (complete board design) and an off-the-shelf printed-circuit-board approach. The reader can make a comparison of the two design complexities.

This text will prepare the reader for reading trade journal articles and other books on the subject to keep up with the state of the art, and further reading references are presented at the end of each chapter.

Donald Baker

LOCAL-AREA NETWORKS WITH FIBER-OPTIC APPLICATIONS

1
INTRODUCTION

Large or long-haul networks, such as telegraph and telephone, have been about since the turn of the century. These networks have well-established design rules because the technology is older and more mature than present local-area networks (LANs). Many smaller networks, such as military command posts and navy shipboard communications, can be considered local-area networks. Many of these networks were nothing more than scaled-down versions of telephone facilities.

With the advent of microprocessors, local-area networks have been designed to handle digital data, video, and voice traffic. LANs are networks with implements located less than 10 kilometers (km) but greater than 1 meter (m) apart. These networks have implements connected within a room, building, campus, and so on.

Networks interconnecting locations within a city or connecting cities within a country are considered long-haul networks. The distances between implements are greater than 10 km and less than 100 km, and networks connecting countries are considered interconnected long-haul networks. Generally, LANs are small networks under the complete control of one organization, with some form of interface when traffic is leaving or entering the facility. In Figure 1–1, two LANs are connected via a long-haul network. LANs 1 and 2 may be installed in buildings such as sales offices and each LAN must use some form of interface to communicate with the standard telephone system, as well as to communicate with each other. The point is, which will be discussed further in the text, that the LANs must have some rules (protocol) if all the terminals wish to communicate with the computers or each other. The rules are decided on by the LAN design. The interface (gateway) must abide by both the telephone company and LAN rules to make communication possible. As one can readily observe, the complexity of LANs can increase rapidly if LANs 1 and 2 have different protocols, which is often the case.

Introduction

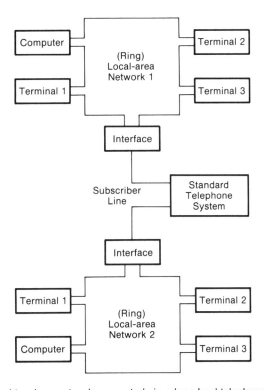

Figure 1-1 A typical two-loop network connected via a long-haul telephone network.

A question one may also ask is why use a LAN at all? A few examples of point-to-point wiring without LAN as compared to local-area networking will make the answer obvious. Examine the network in Figure 1-2(a). This is a point-to-point wired network without LAN capability; any communication is accomplished via a wire path between the devices. The number of connections is expressed in Equation 1-1 for point-to-point wiring.

$$C = \frac{N(N-1)}{2} \quad (1\text{-}1)$$

where C = number of connections
N = number of devices to be connected

For the LAN shown in Figure 1-2(b), the number of connections is equal to the number of devices, which is also the case for Figure 1-1. The two LANs are known as ring and star, respectively, and will be discussed in more detail in Chapters 6 and 7.

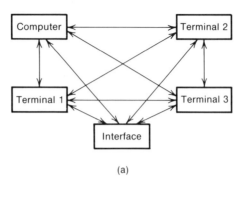

(a)

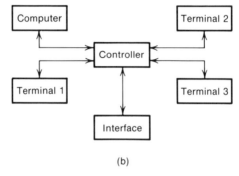

(b)

Figure 1-2 (a) Point-to-point wired network showing connection combinations required. (b) Star-connected local-area network with reduction in connections.

For connecting small numbers of devices, point-to-point wiring may suffice, but suppose 100 terminals and a minicomputer are connected similar to Figure 1–2(a). Then 4950 connections must be made, as compared to 100 for a LAN network. Imagine the task of installing the wiring, which must be shielded for high-speed data transmission. However, the advantages of a LAN will not allow getting something for nothing. The complexities of controlling the network increase with LAN configurations, because as each pair of devices makes a connection the path is shared by other connected pairs. This is commonly known as a virtual connection; that is, the devices communicate as though individually wired point-to-point, but in reality they share a common path.

There are large numbers of applications for LANs, from a simple one- or two-room office to a multiple-building complex. Let us consider a simple application, such as the simple resource-sharing network that can be found in most sales offices. Figure 1–3 is an example.

The central sales office has a disk storage station that contains the sales data base. The data are processed by a minicomputer with a production printer for producing hard copy. This central office services all the satellite sales offices in the region. A system of this type will allow any of the terminals access to the computer and data base. This allows all terminals to share the computing resources, which makes the system highly efficient. Note that the gateway is provided to interface the telephone system with the LAN, as for the previous case in Figure 1-1.

Very often the LAN is an internal system with no access to the telephone system (i.e., no gateway stations). One of the most complex implements in the LAN is the gateway station. If a single manufacturer produces the computer, disk storage station, production printer, and terminals, the interfaces between the LAN and each of these devices are designed as part of the equipment. This is why manufacturers such as Digital Equipment Corporation (DEC), Wang, International Business Machines (IBM), and others have their own LAN. Several other manu-

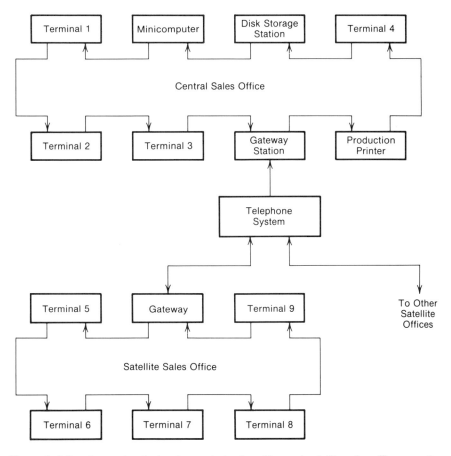

Figure 1-3 Two-loop network showing central sales office and satellite sales office operating through gateways.

facturers may design terminals, computers, and production printers that are DEC, IBM, and Wang compatible. Gateway stations have also been designed that make communication possible between two LANs of different manufacture. As one may observe, the diagrams thus far do not reflect the complexities involved, but as we progress many of these issues will be brought to light.

If the gateway in Figure 1–3 is equipped with a microprocessor or a minicomputer, the central sales office and satellites can be connected in a star configuration. Later in the text a discussion of topologies such as star, ring, and bus systems is undertaken. In this chapter the reader is introduced to some of the problems and solutions available in LANs technology on a quantitive basis.

As one may well imagine, a complete text can be written on gateways. These devices translate the protocol used in one system to that of another. Also, data transmission speeds may be different; therefore, the gateway functions as a transmission-speed translator. With each difference in the LANs, another order of complexity is added to the gateway.

Let us examine what is involved in making two LANs compatible with a gateway. They must be made physically compatible; somehow their electrical or optical connections must be interfaced. The protocols must be translated and, of course, all the transmission facilities must be made compatible. Finally, the software must be compatible. Indeed, this seems like a monumental task and quite often is, but if a methodical approach is used to describe networks, some of the complexities can be removed. This is one of the reasons for describing LANs using the International Standards Organization (ISO) model. This model is the topic of discussion in Chapter 2.

The text will chiefly consider data LANs, with brief introduction to digital voice and video. Digital telephone has some problems that limit its use on data channels, which is also the case for digital video.

Another use for LANs is electronic mail. At times when long-distance rates are favorable, such as weekends or after 5:00 P.M., message traffic can be conveyed between business establishments automatically. When the business day begins, a local data base can be interrogated for messages. If sales forms are reduced by facsimile machines to digital form, they may be transmitted in a similar fashion, with the facsimile as a LAN implement.

A large fast-food chain uses a unique technique for changing prices on food items. The pricing changes are sent, during cost-favorable telephone-rate periods, to the restaurants. The prices stored in the cash registers are changed during the night. When the restaurant is opened the next day, the prices of all items reflect the changes. These price changes were performed by a serviceperson previously. Now it is all accomplished by a computer and LAN in a matter of hours, rather than days.

The computer- and telephone-connected LAN can also maintain automatic inventory control. The manager of such facilities is relieved of this major responsibility by the computer.

One of the most challenging applications for LANs is automated production facilities, such as needed by the auto industry. The LAN must function in a harsh environment (i.e., large temperature excursions, large electrical transients, and dust). In applications of this type, the terminal operator will not be a skilled programmer; therefore, the application program will require menus and prompting. The terminals can also be equipped with speech synthesizers to improve operator interaction with terminals; as speech-recognition technology becomes more mature, the terminal may not require keyboards.

There are a number of goals that local-area networks will fulfill. The LAN connecting computers together at various locations within companies allows for more efficient use of these resources. Also, the failure of one computer may slow down the facilities, but operation may continue at a reduced performance. However, system reliability is increased because of redundancy in the system. Some organizations cannot tolerate catastrophic failure; therefore, expensive backup computers must be maintained if a LAN is not implemented. Examples of these facilities are military, industrial process control, and banking.

Distributing computing power with LANs is possible because of relatively inexpensive computers and communication networks, which have been dropping in cost since 1970. Communication facilities cost more than computers, but the situation was reversed prior to 1970.

Most mainframe computers are only a factor of 10 faster than the largest single-chip microprocessor-based computer. But the mainframe cost is well over a thousand times more expensive, which indicates that the price–performance ratio of microprocessor-based systems is superior.

Several microprocessor manufacturers are making 32-bit data-bus machines. These can either be paralleled for array processing or operated with coprocessors to increase processing power. Many floating-point processors are appearing in printed-circuit-board form; these allow the microprocessor computer user to upgrade systems.

Processing limits LANs operations, but not the networks themselves. The cable used for interconnecting can be bandwidth limited, attenuation limited, or both. Fiber-optic cable plants are one solution to some of the problems. Fiber optics offer wide bandwidth and low loss per kilometer. The telephone companies are heavily involved in this technology, which will ensure its steady growth. One advantage of fiber-optic cable is its relative security when compared to copper. The usual methods of tapping lines will not suffice. This topic will be addressed in more detail in Chapter 5.

This chapter has given the reader an overview of the direction in which industry is moving with respect to LAN technology.

2
INTERNATIONAL STANDARDS ORGANIZATION NETWORK MODEL

Networks are designed in a highly structured manner to reduce their design complexity. Before discussing the various layers in the International Standards Organization (ISO) model, some background on the physical structure of serial data links will be given.

Point-to-point connections are those with copper or fiber-optic cables between the communicating nodes, such as in Figure 1–2(a). We have a virtual connection when multiple channels of data are sent over the same copper or fiber-optic path. Most of this text will deal with virtual connection rather than connecting each implement point-to-point, as in Figure 1–2(a).

Two popular methods of forming virtual connections are shown in Figure 2–1. Figure 2–1(a) is called time-division multiplexing (TDM); this is an oversimplified diagram. If the data in terminal 1 are transmitted to terminal 5, one character at a time, and the multiplexer and demultiplexer form the path, transmission can occur for one character provided they are connected during the same time slot, such as TS1. Furthermore, suppose terminals 2 and 4 are also communicating one character at a time; the multiplexer and demultiplexer will connect the two units only during their specific time slot, such as TS2. These two connections, as observed by the terminal operators, will be simultaneous. The transmission paths are virtual connections using TS1 through TS3 as time-slot channels.

Figure 2–1(b) shows the other technique, known as frequency-division multiplexing. This diagram again is oversimplified to illustrate a point. When a terminal wishes to communicate, it sends out its address and the terminal it wishes to communicate with on channel f_0, using frequency-shift keying codes with frequencies as follows:

$$f_0 - f_M = 1 \text{ binary} \quad \text{(M)}$$
$$f_0 + f_S = 0 \text{ binary} \quad \text{(S)}$$

International Standards Organization Network Model

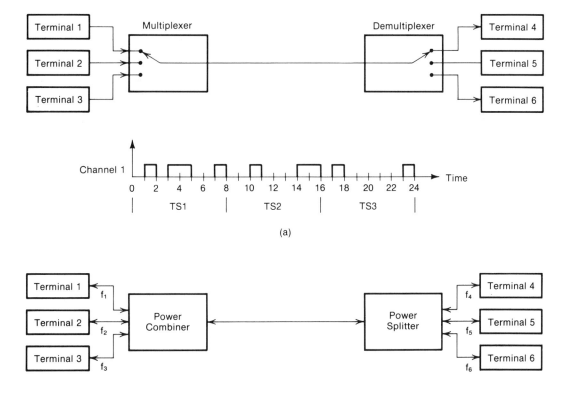

Figure 2-1 (a) A simplified time-division multiplexed (TDM) network. (b) A simplified frequency-division multiplexed (FDM) network.

All terminals monitor f_0 when they are not communicating. The answering terminal will tune its receiver to the transmitting terminal (e.g., channel f_1 for terminal 1). The transmitting terminal will then tune its receiver to the answering terminals, for example, frequency f_6 for terminal 6.

Thus a communication path is established between terminals 1 and 6. Terminal 1 transmits on f_1 and receives on f_6, while terminal 6 transmits on f_6 and receives on f_1. Note that this technique allows full-duplex communication (bidirectional), and transmission is simultaneous.

This second technique is similar to citizen-band radio (i.e., channel 9 is

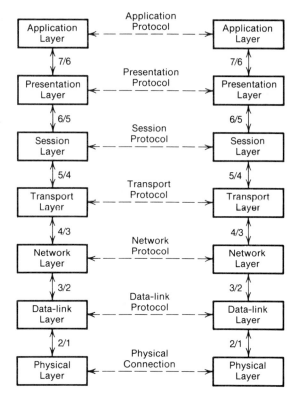

Figure 2-2 International Standards Organization (ISO) layered network model architecture.

monitored). When a person wishes to call, he or she establishes initial contact through channel 9 and switches to another channel to communicate.

Often combinations of both techniques are used in LANs. Each technique has trade-offs to be investigated. A further discussion is reserved for Chapter 5.

The ISO model is segmented into seven layers, as shown in Figure 2-2. The layered approach to LANs can best be described by the following principles:

1. Each level should perform a well-defined function.
2. The layers should comply with well-recognized standards if possible.
3. Layer interfaces should be well defined and should be chosen to minimize information flow across interfaces.
4. The layer size should not be too large because that would make it unmanageable. Also, it should not be too small because that adds unnecessary complexity to the model.

The layers connected with dashed lines in Figure 2-2 are virtual connections (i.e., each layer communicates as though it is physically connected). Only layer 1 is physically connected as shown in the diagram. Each layer has an interface between the layer above and the layer below it. Let us now examine each layer to determine its function.

Physical Layer

The physical layer establishes the physical connection. The design issues consider first the type of cable plant to implement the design (i.e., fiber optic or copper). Implementation of error-checking hardware is necessary to ensure that bit patterns have no errors. Note the reference to bit patterns, not messages, because message length is of no consequence at this layer. Connectors, types of transmitters and receivers, lightning protection such as avalanche devices, electrical compatibility, noise, and of course optical compatibility for fiber-optic cable plants are all physical layer design issues.

For multiple physical connections, this hardware layer may look like a single connection to the layers above. Layers above may also provide this service (i.e., a single connection may be multiplexed to look like several connections). The physical layer will also control whether the channel is full or half-duplex (i.e., bidirectional or unidirectional at any one time).

This layer is also concerned with mechanical issues, such as the form factor of network interfaces. The design of this layer is performed by electrical and mechanical engineers.

Collision detection is also a task of the physical layer; this will be discussed in more detail in Chapter 5.

Data-link Layer

The data are assembled into manageable groups of bytes called frames. Note that the beginning and end of the frame must be known. The flags at each end must be distinguishable from data; for long messages, multiple frames must be sent. This layer adds the necessary sequence numbers to the frames to allow data with multiple frames to be reassembled into their original state at the receiving device.

The flags must be removed when the data are received; this makes the data-link layer transparent to the layer above (network layer). If a frame is damaged during transmission, this layer will initiate retransmission of the damaged frame. Also, when frames require an acknowledge (ACK) or not acknowledge (NACK) response, the data-link layer will send these messages.

Error-checking algorithms are sometimes employed at this level because burst errors can go undetected for certain types of physical-layer cyclic redundancy checking (CRC). The commonly used CRC will detect burst errors with reduced accuracy when these errors are greater than 18 bits with a 16 bit CRC register. The data-link layer is also responsible for preventing receiver overflows (i.e., the transmitter is kept informed of the receiver buffer status during transmission).

Data management is provided during full-duplex and half-duplex operation. Often ACK/NACK messages are piggybacked on message traffic in lieu of sending a special ACK/NACK frame. This piggybacking technique will be discussed in more detail in Chapter 6.

Network Layer

The fundamental purpose for the network layer is to establish a virtual circuit and maintain it during transmission. The communication between network nodes is in the form of packets. They contain the addressing and supervisory information necessary for network operations. The packet will be discussed in more detail in Chapter 7.

Another task of the network layer is interfacing with the next layer in the hierarchy, the transport layer. The network layer will assemble and disassemble transport messages during transmission and reception, respectively.

Traffic or flow control is another function of this layer. The traffic flow may be such as to cause congestion in the network; this layer will regulate and keep traffic flowing in an orderly manner.

When traffic is monitored for accounting purposes, such as is the case for a larger network, the network layer will perform the necessary tasks. The network layer may monitor bits, characters, packets, or tune to correct channel frequency when customers use the facilities, whatever is necessary to produce billing. Also, different nodes may have different accounting methods and rates to be included in billing. As one may well surmise, this is a very important layer.

An additional responsibility of this layer is maintaining priorities in a prioritized system (i.e., keeping messages in proper order). For example, part 1 of a message must arrive before part 2, and not vice versa. The routes are usually determined through the use of routing tables. These are usually hard wired into the network for small systems, whereas larger networks may establish routing at connect-in to assist with traffic conditions.

The network layer may also be tasked to keep peer layers informed about traffic status. This may, for example, result in a readout indicating network status, which the operator may monitor. The network layer is usually implemented as part of the input–output operating system drivers.

Transport Layer

The transport layer takes session layer data, the next higher layer in the ISO model, and breaks them up into smaller units. The transport layer multiplexes these data and puts them onto the network. The transport layer can also present the session data to more than one network connection to improve throughout. This is basically a multiplexing layer.

The transport layer is also responsible for end-to-end error detection and recovery, and it monitors the quality of service. This layer addresses the end user without regard to route or address of machines along the route between end-user machines.

Transport-layer multiplexing involves disassembling (transmitting) and assembling (receiving) of session layers. This layer also will maintain session-layer segments in their correct order during transmission and reception. This layer isolates the session layer, which can be part of the host-machine operating system, from changes in hardware.

If cost of service is expensive, this layer may also multiplex several sessions onto a single network connection. This type of situation may also be encountered during heavy traffic flow to relieve congestion. The fast host machine must be prevented from overrunning the slower machine. This layer will be responsible for this function at the session level.

Session Layer

The session layer establishes the connection between two users, which is actually the end result of the whole process. More technically, the connection is established between the presentation layers, which perform certain transformations on the data before they are transferred to the application layer. By providing the necessary synchronization between end-user tasks, it sets up communication options such as whether the connection will be half- or full-duplex.

The session layer provides the necessary addressing to be implemented by the users or their programs. This layer will map addresses to names (i.e., it performs a directory function). As one may readily observe, this is a function of most host operating systems. This layer provides smooth operation between users (i.e., opening and closing of data transfer between users). It also provides file transfer checkpoints.

The session layer provides the necessary buffering between two end users. Such buffering may be necessary before the session data are conveyed to the end user or before data can be transmitted.

Presentation Layer

The presentation layer translates or interprets the session-layer data for the application layer. When computers are communicating and the application layers are to understand the information transferred by two dissimilar machines, they must use a common syntax to represent alphanumerics, file formats, data types, and character codes. The presentation layer negotiates the syntax (often called the transfer syntax). This allows the two machine application layers to communicate.

The transfer syntax allows the originating and destination application process to be of a different format and the exchange of information is still possible. To negotiate, the application layer must supply the presentation layer with sufficient information about data to be exchanged and rules for the negotiation procedure.

The application layer need not provide specific information, such as file formats, but it need only supply what type: financial, display, and so on. The originating presentation layer must provide a transfer syntax that both are capable of using to represent the information. The two communicating entities must agree on the transfer syntax or communication cannot proceed. If local mapping between the originating and transfer syntax can be performed, then the two applications can communicate, but this is not a communications issue. The ISO does not standardize on all possible syntaxes because, as one may well imagine, this would be a lifetime's work for some committee. The ISO recommends some of the more popular syntaxes. The ISO is, however, standardizing the negotiating protocols for the transfer syntax. More of these issues will be discussed later in Chapter 7.

Application Layer

The application layer is the layer that the end user is acquainted with. It is the layer that allows the user access to the computer. The application layer is responsible for making all other layers transparent to the operator of the equipment. This layer provides such services as removing line feeds, carriage returns, and control characters from the text; it also inserts appropriate subscripts, tabs, and the like, into the text.

The application layer may provide such service as log in and password checking. This type of service might be needed in banking or time sharing of computer services. Also, one would need this type of service to maintain network security, which is of the utmost importance with the advent of computer hackers (an expression used to describe individuals involved in breaking into secure networks). This of course is a problem to the banking industry with the increased use of automatic banking.

The application program is responsible for setting up through negotiations an agreement on semantics, such as the case of a bank communicating with a loan company where they must agree on the information to be transferred (e.g., the transaction deals with credit references rather than money transactions). During a communication period, the application layer may be capable of switching between semantics, commonly called context switching. An example would be two communicating partners transferring credit checking and then switching the context to information transfer to airline reservations. The ISO has agreed on standardization of service protocols such as management, file transfer, and virtual terminal types. A further discussion is reserved for Chapter 11.

TABLE 2–1
Standards

Sponsor Organization	Affiliation	Membership	Influence in Marketplace
International Standards Organization (ISO)	Voluntary compliance	Participating nations with U.S. representation	Close relationship with CCITT
International Consultative Committee on Telegraph and Telephone (CCITT)	International Telecommunications Union (a United Nations Treaty organization)	Most of western Europe and with U.S. representation	Enforced by law for countries that have nationalized communication
European Computer Manufacturing Association (ECMA)	Computer suppliers selling in western Europe	Trade organization of suppliers	Contributes to ISO and issues ECMA's standards
American National Standards Institute (ANSI)	Voluntary compliance	Manufacturers, users, communications companies, and other organizations	U.S. representation in ISO (International Standards Organization)
National Bureau of Standards (NBS)	Government agency	U.S. government network users DoD's ARPANET implementers	Federal information-processing standards purchased by DoD need not comply
Institute of Electrical and Electronic Engineers (IEEE)	Professional society	Society members	Contributes to ANSI and issues standards: IEEE 802, IEEE 488
Electronic Industry Association (EIA)	U.S. trade organization	Manufacturers	Contributer to ANSI physical layer, RS422 RS232-C
Department of Defense (DoD)	Government agency	Government and military	Military establishment
Others	Voluntary de facto standards	Various manufacturers	Standards for own equipment

The ISO standards are being widely accepted but by no means are they the only standards. General standards are presented in Table 2–1. As one can observe, certain standards are compulsory for communications, for example, throughout a country, or for government equipment. In time of war or national emergency, components within systems must be replaced easily during equipment failure. The standards presented in the table are the most popular, but others are generated by manufacturers and organizations.

Most of these standards are not similar. Therefore, the network designer must be careful when designing interfaces for standards (i.e., the standard may not be well recognized). If this is the case, the product will have a limited lifetime and market.

3
COMMON NETWORKS

The contents of this chapter will aquaint the reader with a cross section of local-area networks. The first is ARPANET, which is the earliest design of a packet-switching network. This architecture was pioneered by the Advanced Research Projects Agency (ARPA), which provided a test bed for packet-switched-type networks such as LANs.

The next three networks are baseband LANs, which have certain subtleties that will be investigated. These three also represent a good share of the LAN market. Wang and Sytek networks represent broadband local-area networks that are very useful for transmission of diversified data. The X.25 is an important standard for the first three layers of the ISO model. The interface is for public networks.

Packet Switching

Packet switching was introduced by ARPANET. Messages may be broken up into manageable segments and sent when the total message is complete. The transmission is a burst of data, and network control functions process packets in an efficient and timely manner. Packet switching is primarily used on data-communication networks, but digital voice has made the possibility of packet-switched voice more feasible.

One may ask, why a packet-switched telephone? Telephone plant facilities are not used very efficiently. As an example, the average telephone call lasts approximately 3 minutes, and when circuits are used they generally operate half-duplex (i.e., only one person talks at a time). The use of the voice circuit is 10

TABLE 3–1
Comparison of the ISO Model with Other Networks

Layer	ISO (Model)	ARPANET	SNA	DECNET	ETHERNET
7	Application	User	End user	Application layer	UNET mail protocol
6	Presentation	Telenet interactive terminal interface parameters, FTP (File Transfer Processor)	NAU service (network, addressable units)	Application layer	Unet file and terminal protocol
5	Session	(None)	Data flow control	(None)	(None)
4	Transport	Host to host / Source to destination IMP (Interface Message Processor)	Transmission control	Network services	UNET transmission control protocol
3	Network	Source to destination IMP (Interface Message Processor)	Path control	Transport	UNET INTERNET protocol
2	Data link	IMP to IMP	Data link control	Data link control	ETHERNET controller
1	Physical (baseband)	Physical (baseband)	Physical (baseband)	Physical (baseband)	Transceiver and cable (baseband)

percent during peak load periods. As one can readily see, packet voice will promote more efficient use of plant facilities and prevent possible tariff increases.

ARPANET

ARPANET provides host-to-host protocols for data exchange between machines. The specifications of format and procedure for data exchange are well documented; the user protocol is only functionally specified. The functions are open, close, receive, send, and interrupt servicing. This allows user tailoring of interfaces, which makes possible the use of other computers on the net if they comply with host-to-host protocols. This has established important principles for protocol specifications.

Comparing ARPANET with the ISO layer (see Table 3–1), we see that it has a physical layer, but the second layer is a combination of layer 2 and part of layer 3 with reference to the ISO model. The ARPANET third layer is composed of part of the ISO network and transport layer, and its fourth layer is part of the ISO transport layer. ARPANET has no session layer, but the sixth and seventh layers have no ISO equivalent. See Table 3–1 for a summary. This summary will be referred to later when protocols are discussed. The interactive terminal parameters are discussed in Chapter 10 along with virtual terminal protocols.

IBM's Systems Network Architecture: SNA

The SNA was devised by IBM to allow large customers such as banks, loan companies, manufacturers, and other businesses to integrate their data-processing facilities. In this manner, the facilities can be used much more efficiently; if computer failures occur, some backup computing power is available at a reduced performance. As an example, a bank has two mainframes with 50 terminals located throughout the main banking facility and several branches. All 50 terminals can access either computer via a LAN. If a computer fails, the bank will have one computer that may be overloaded with the 50 terminals. The terminal operators will only note that terminals are reacting more slowly to processing traffic.

Prior to the advent of local-area networks, most products and interfaces had protocols that were unique to the product line. IBM had hundreds of communication products with a few dozen access methods and a dozen or more protocols. Imagine, then, the chaos that existed when a network of a sort was configured, each network being a new engineering challenge with no doubt new interfaces and protocol modifications. It must be rather refreshing to have all components of a network drop shipped and integrated at the premises with only minor modifications necessary.

Interfaces and protocols allow for change if the LAN is to remain viable. Innovations that increase performance and expand facilities are required if the network is to succeed. The SNA has these attributes.

SNA local area networks consist of nodes that are machine types (see Table 3–2). Observe that microcomputer terminals with preprocessing capability incorporate several of the node attributes into one machine. Many of the supermicrocomputers have the computing power of a minicomputer.

TABLE 3–2

Node Type	Description	Comment
1	Terminals	
2	Controllers	Supervisor peripherals
3	Undefined	
4	Front-end processors	These preprocess data
5	Host processor	Mainframe computer usually

Figure 3–1 depicts a two-domain SNA network; this can be extended to larger networks. The domains are managed by the systems services control point (SSCP). There is one for each type 5 node; it monitors and supervises the domain. At least one or more logical units (LU) are required for user processes. The node is provided with a physical unit (PU), which is used by the network to perform administrative functions such as testing and putting the node on or off line.

The first two SNA layers function similarly to the ISO model. Layer two of

Common Networks

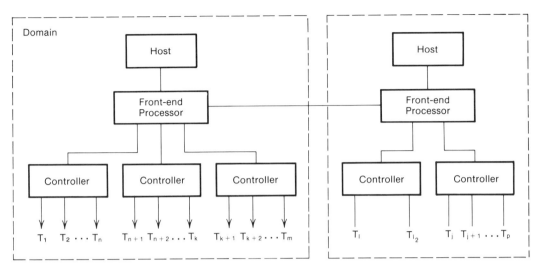

Figure 3–1 A two-domain IBM System Network Architecture (SNA).

the SNA structure closely resembles the synchronous data link control (SDLC) protocol, which is produced in integrated-circuit form, such as the WD1933, and manufactured by Western Digital.

The path control layer, as shown in Table 3–1, performs the networking and part of the transport function. This layer deals with routing and congestion-control issues as layer 3 of the ISO model, but it also performs several layers of multiplexing and demultiplexing. The path control (PC) delivers data from origin to destination transmission control (TC) layers. Each link queue contains all the data destined for a particular adjacent node; this is the primary station when the queue is served by a secondary station, and vice versa. Unrelated packets may be associated into it to enhance transmission efficiency, and it can cope with hierarchical SNA addressing.

The next layer is transmission control. This layer provides the necessary transport service, and when the transmission is complete it will delete sessions. Session is not the same in SNA as in the ISO model. Once the service is established it manages message priorities, regulates the rate of data flow between processes, provides buffer allocation, provides interface to upper layers, and performs, when required, encryption and decryption. Note that one must be careful when reading about SNA not to get session confused with the ISO terminology. The subnet properties are also kept transparent to the peer layers by the transmission control function (the subnet serves as a connection between hosts).

Data flow control, as one can observe in Table 3–1, supervises the session layer (SNA type); that is, it determines the session flow direction, origin to destination or destination to origin. This level also provides error recovery. Header information normally included in the layer of the ISO equivalent is embedded in the transmission control layer.

The sixth SNA layer provides presentation services similar to the ISO layer and session services (not the same as ISO) for establishing a connection. Additional network services control the overall network.

The end-user layer is self-explanatory. It corresponds closely to the ISO application layer.

Digital Equipment Corporation Network: DECNET

DECNET was designed to allow DEC's customers to interconnect computers into a private LAN. Previously, the only method DEC computers had for communication was hardwired point-to-point (i.e., if they were physically connected). Now they may communicate through intermediate nodes (e.g., machines A and B are connected to C). Previously, A could not communicate with B through C, but with DECNET the nodes only need be on the network to communicate.

A comparison of DECNET with the ISO model is provided in Table 3–1. Note that the transport and network layers are interchanged when compared to the ISO model and DECNET has only five layers. Also note that there is no session layer. The layers will now be examined to determine their functionality.

The DECNET physical layer is similar to the other architectures previously discussed. The data-link layer has some differences in frame construction. It is byte oriented. A discussion of this is deferred to Chapter 6. The DECNET message packets may be routed independently through the subnet, whereas some LANs require the same route for all the packets of a message, typically SNA networks. DECNET has no session layer embedded in other layers as is the case for SNA networks. Transformation of data in DECNET must be performed indirectly; that is, encryption, decryption, text compression, and various code conversions must be accomplished with remote files.

Xerox Corporation: Ethernet

Ethernet is a baseband network dedicated to LAN. The physical part consists of a 50-ohm cable with an adjacent transceiver, which is connected to the data-link layer controller (Ethernet controller) on the backplane of the equipment. This latter connection is made through the use of an interface cable; it supplies power, receives and transmits, and monitors collision detection signals. A simplified block diagram of a typical Ethernet connection is shown in Figure 3–2. Note that signal isolation is required because the system must have a single grounding point (i.e., each station must not be grounded, similar to a CATV application). This makes the coaxial cable an antenna, which is responsible for excessive amounts of electromagnetic radiation. This is an ideal application for fiber optics, which will be discussed more thoroughly in Chapter 5.

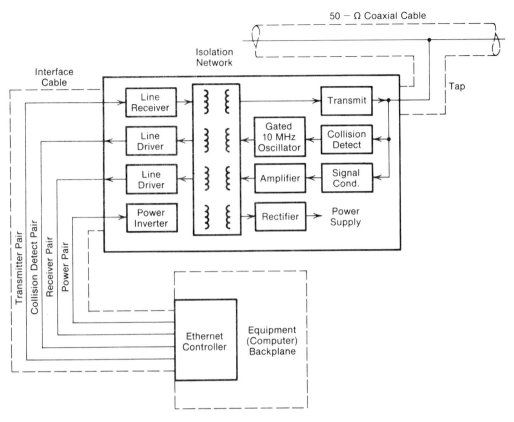

Figure 3-2 Typical Ethernet connected station.

Ethernet standards and specifications cover the first two layers of the ISO local-area network model. In the model shown in Table 3–1, the layers above the Ethernet controller are implemented in UNET. The internet protocol (IP) and transmission control protocol (TCP) are both implemented by using U.S. Department of Defense standards. These are examples; actually, these upper layers are defined by the customer or user. The upper-layer UNET file transfer program is a technique for transfer of UNIX files between two host machines. Again, this layer is only superficially covered because it can be defined by other than the UNET file and terminal protocol. The upper-layer UNET standards allow direct access of UNET internet protocol, which is not possible with the other models discussed.

Ethernet has been implemented in several DEC installations by systems integrators. The DEC machines include PDP-11 and some VAX series computers, with the controllers designed using LSI-11s. The UNIX operating system has been used extensively for these applications. Ethernet has an inherent shortcoming, the maximum end-to-end length is limited to 2.5 km. This particular dilemma can be solved using fiber optics. The increase can be rather dramatic (i.e., the length limit can be increased by a magnitude). The increase in end-to-end length due to fiber optics is highly dependent on the wavelength used; 1300 nanometers (nm) is preferable with LED transmitters in conjunction with PIN FET receivers. These issues will be covered more thoroughly in Chapter 5. For larger distances, lasers and monomode transmission techniques are more suitable, but monomode networks are expensive.

There are a large number of other baseband networks in the LAN market. Many other manufacturers of LANs have incorporated features from the networks presented here. By studying these networks, the reader can more appreciate their design subtleties.

Wangnet

Wangnet is not included in Table 3–1 because it is a broadband network. Wangnet will manage rather diverse forms of data and it can handle these data simultaneously. The cable may have frequency-division multiplexed (FDM) channels with video, audio, or digital-type data. As one can readily observe, the LAN finds great use in office environments. When one makes a decision to implement a broadband LAN, certain considerations must be examined. The basic questions are whether a simple baseband system will suffice because the data being manipulated are all digital and low data transmission rates are adequate.

One of the advantages of Wangnet is the use of CATV technology, which has been in use for 20 years. Radio-frequency (RF) components are highly reliable because of the mature technology. Also, the cost of CATV components is less than that of many baseband LANs. The bandwidth of Wangnet is 340 megahertz (MHz) with the bandwidth of 42 MHz reserved for user RF devices. This band is located from 174 MHz to 216 MHz (channel 7 to 13 in the TV band).

Wangnet consists of a number of communication services, which are labeled Wangband, interconnect bands, utility band, and peripheral attachment service. The diagram in Figure 3–3 shows the band allocations. The 10- to 12-MHz and 12- to 22-MHz bands are point-to-point network components with fixed-frequency modems to provide the virtual connections. The band from 48 to 82 MHz is controlled by a data switch, which is controlled by a computer. Its functions are similar to a miniature telephone exchange. A call is established when the data switch selects an unused frequency and initiates the tuning of a frequency agile modem to establish the channel.

A utility band is provided for user RF devices; for example, TV channels 7 to 13 on the standard broadcast band may be provided on the cable by the user. Other channels may be broadcast via this cable if they are mixed down or up to this band.

Perhaps the most important band is the CSMA/CD (carrier sense multiple access/collision detection) Wangband. This FDM channel has a fairly high transmission speed (12 Mbits/second) and it provides LAN service. The Wangnet pro-

Common Networks

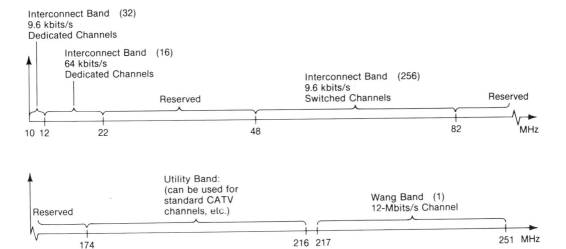

Figure 3-3 Description of the Wangnet frequency allocation.

tocols emulate the lower five layers of the ISO model (i.e., physical to session layers).

Propagation delays will be examined in Chapter 5 to determine the impact on LAN facilities. The packet size also to be discussed will be dependent on the maximum propagation delay. In addition, a comparison will be made between fiber-optic and copper cable.

LocalNet

LocalNet is a broadband network manufactured by Sytek, Inc. The RF coaxial cable used for implementing this network is a standard CATV type with a 5- to 400-MHz bandwidth.

The broadband frequency allocation is a midsplit system. Outward signals from the head end are in the 226- to 262-MHz band and inward signals to the head end are in the 70- to 106-MHz band. When a terminal conveys data to the computer (see Figure 3–4), the digital data are passed to the PCU (packet communication unit) via an RS232 link. The PCU's modem will impress the data on the RF carrier with frequency-shift keying (FSK). The data are now in RF form. They will pass through the 50/50 head end to be upconverted in frequency to the 226- to 262-MHz band and will be detected by the receiving PCU and presented to the computer via a RS232 link. Note that when the computer answers, the outward signal must pass through the two amplifiers and to the head end to be upconverted to the receive frequency band. This is a highly simplified version of how the equipment actually functions.

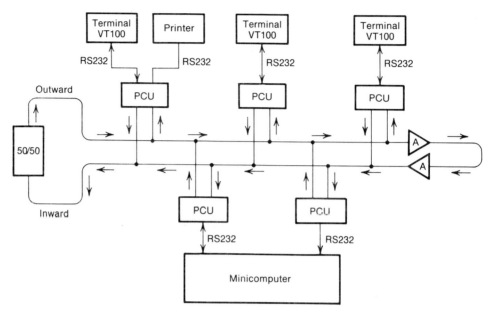

Figure 3-4 A small Sytek LocalNet installation with four network implements connected to four minicomputer RTS232 ports.

Each of the bands has 120 (300 kilohertz, kHz) channels. The modems are arranged in groups by letters A to F and each of the letters has 20 channels per group. These lettered groups have a 20 × 300 kHz = 6-MHz bandwidth. Each channel within the group operates at 128 Kbits/s using the CSMA/CD protocol. With time-division multiplexing of each of the 128-Kbit/s channels, 200 or more terminals can be serviced by a single channel. T-bridges can be used to implement more than 20 channels. If bridges are used, many thousands of terminals may be serviced. Thus the network can easily be expanded to meet most user needs.

LocalNet 20 has a transmission rate on the cable of 1 Mbit/s; this system is useful for low-duty-cycle users such as terminals connected to minicomputers. The LAN will cover up to a 50-km geographic radius.

LocalNet 40 is a higher-transmission-rate system. It is useful for higher-duty-cycle users and operates at 2 Mbits/sec. Both LocalNet 20 and 40 are unsuited for intercomputer transfers.

Sytek has several other implements that attach to their cable. A network control center (NCC) is available that serves a multitude of functions. This equipment has a network monitor that statistically monitors the network channels and records traffic density. It also can be used to forward and store electronic mail (i.e., the station to receive the mail will be notified and it can interrogate the NCC). This device can store a directory routine that allows users to use names that can be translated into terminal addresses.

The NCC hardware consists of a Motorola MC68000-based microcomputer with a streaming tape drive (for the operating system) and 20-megabyte Winchester disk drive. The connection to the broadband cable is via two LocalNet 200/100 PCUs, which have eight RS232 ports each. Hardware will be discussed in more detail in Chapter 5, Example 5.1. This equipment has a UNIX-based operating system, which increases its utility.

The 200/100 LocalNet PCU follows the ISO model with some exceptions as to the functionality of each layer. For example, the physical layer has an RF coaxial cable and RF agile modem. The digital transmitter must frequency-shift key the modems and transmit modulated carrier, while incoming data must be RF detected at the receiver and converted to digital form. Other differences will be discussed as they are addressed in examples in Chapters 5 through 11.

Note that we have only discussed the Sytek band on the cable, which is only 36 MHz in each band. Thus there is a great deal of spectrum remaining for future expansion of facilities, such as CATV channels, audio channels, facsimile, and other user services, as well as Sytek-related attachment of band implements. Some experimental digital telephone facilities have been successfully used in this type of system.

Commercially available remoting has been accomplished using fiber-optics modems that operate to 5 km. These modems consist of an electrical-to-optical conversion at the RF cable drop and at the far end (remote location) an optical-to-electrical conversion. At the time this book was being written, only a transmit remote was commercially available at 70 MHz. But the author expects a receive modem to be available by the time of publication. Also, advances in fiber optics can be expected to extend the remote distance to a minimum of 20 to 30 km. This technology is available at present. The fiber-optic remote would be useful where excessive amounts of electrical noise are present, such as near mill machinery, punch presses, and high-voltage switch yards and to some extent where radiation is present. The fiber-optic cable may be completely nonconductive; therefore it may be routed in the same cable runs as electrical power cable with no adverse effects. In Chapter 5, several commercially available fiber-optics components will be discussed.

X.25

The final discussion of this chapter will address X.25, which is by no means the least important. It is more commonly referred to as CCITT recommendations X.25 (for public packet-switched networks). This is a standard being adopted by many countries, including the United States. It standardizes some of the LANs' physical, data-link, and network layers of the ISO model. The X.25 standards are similar, but there are some subtle differences, which will be examined in the appropriate section of the text. X.25 provides a virtual connection between computers, which of course is only half the job of communication.

Software compatibility is the other half of the job. UNIX operating systems will help to alleviate some of the headaches of communications engineers, but many manufacturers who tout UNIX operating systems have in reality UNIX-like operating systems. The conflict lies in that the communication between the computers almost operates correctly. The standards are meant eventually to eliminate all loopholes and make the equipment completely compatible. They have both good and bad effects. The good result is that equipment may be purchased anywhere and it will operate or interface correctly. The negative aspect is that a better design may be found for interfacing, but cannot be used because of the standards. One of the most successful organizations is the U.S. Military and their MIL standards.

This chapter has supplied the reader with an overview of several local-area networks that are very common throughout industry. The text should have posed several questions to the reader, which we will endeavor to answer throughout the next eight chapters. The questions are:

Review Problems

1. How are physical connections implemented?
2. How does fiber optics enter the physical scheme?
3. What are some of the common protocols and are they available in integrated-circuit form?
4. What are some of the common network techniques?
5. How can I use all this LAN information effectively in an actual design?
6. What are some of the performance trade-offs?
7. When is it preferable to use a broadband over a baseband network?

4
NETWORK TOPOLOGY

In this chapter we will investigate techniques for the analysis and design of network topology. Much of the information presented will be used in Chapter 7 for design and analysis. Before topologies are discussed, one needs to determine when broadband or baseband is more advantageous. This decision will affect the topology and may simplify the analysis to some extent.

Let us first examine broadband networks. One important advantage of the broadband FDM techniques is the network's ability to handle analog and digital information. Also, the information can be processed simultaneously. Coaxial cable, RF amplifiers, cable drops, RF modems, and other RF components have been used in the CATV industry and have a high level of reliability. Broadband networks may have as a subset many TDM channels equivalent to several baseband cables. Broadband facilities will allow the user to upgrade transmission rates of the TDM channels with little or no modifications to the RF components. Therefore, future upgrades in facilities may require modem modifications and protocol changes but no changes in the cable plant, which of course could be quite costly. These cable plants can be easily expanded.

Broadband networks also have a negative side in that these facilities are implemented with RF modems, which are more complex and expensive than baseband transmitters and receivers.

Fiber-optic cable plants are not suitable for broadband operation, but this may change in the near future as the technology becomes more mature. A problem with present-day fiber optics is that the total broadband spectrum must AM modulate a single carrier wavelength, which requires two levels of demodulations at least. Also, the AM modulation index cannot be much greater than 50 percent or distortion results, and when laser diode transmitters are used, they can be easily damaged. Many transmitter manufacturers incorporate limiters to protect laser sources.

The broadband cable bus uses some sort of a head-end unit which can cause a catastrophic failure when it has a malfunction. Sytek, for example, has a dual unit that switches in a backup head-end frequency converter and sounds an alarm when a head-end failure occurs. But this increases the cost of plant facilities.

Baseband LANs are less complex than broadband types. The transmitters and receivers lend themselves well to monolithic or hybrid circuit techniques, which will keep cost low. Some baseband transceivers are actually repeaters. This implies that all the machines must remain powered up unless the network outlet is equipped with the transceiver. For this particular situation, fiber optics can be used to implement the cable plant, and each repeater station is equipped with relays to allow the outlet to function as a repeater when the equipment is disconnected from the outlet. The outlet cost in small quantities is less than $100 per outlet. Other baseband networks use restrictive taps on the cable, which are much simpler to implement, but repeaters can be required periodically as needed.

When making a decision on whether to implement a LAN in broadband or baseband technology, one must first estimate the diversity of the data. As an example, if standard TV broadcasts are to be distributed by the LAN, the decision is already made (broadband is the only choice). If, however, teleconferencing is to be implemented using slow-scan TV, perhaps baseband techniques can be used successfully because 9600-bit/s slow-scan TV is readily available. If terminals are going to be connected to a minicomputer and they are only used for interrogating and upgrading the data base, a baseband LAN would be adequate and the most cost effective. But in a situation where computers are interconnected, broadband transmission can be advantageous.

Network Topology

First we must examine some of the more common topologies, such as the bus, ring, and star, that are commonly implemented in LANs. Also, there are situations in which combinations of the three are encountered.

A bus configuration is shown in Figure 4–1(a). This configuration consists of a passive cable such as an Ethernet bus. The drops are connected to the network implements, such as terminals, minicomputers, disk storage stations, and gateways. If the bus is broken or disconnected between network implements, such as between 3 and 4 in the figure, the network will function as two buses. Many installations require a backup cable installed to prevent failure of the system if a cable is damaged. Another technique is to install accessible connectors at strategic locations throughout the installation and provide long jumpers to circumvent the damage. This can be done automatically for situations where high reliability during emergencies is required, situations such as natural disasters or tactical operations. The complexity of the backup switch service will be complex and costly, however. A fiber-optic bus can also be implemented. Instead of cable taps, couplers are needed. A more detailed examination will be conducted in Chapter 5.

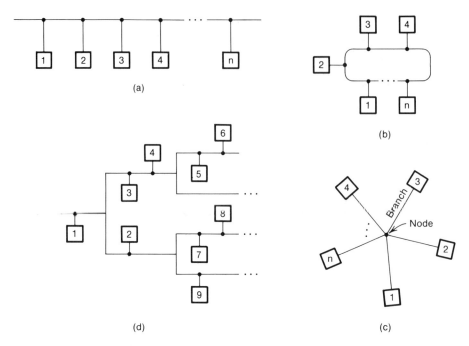

Figure 4–1 Typical topologies of local-area networks. (a) Bus network. (b) Ring network. (c) Star network. (d) Tree network.

The ring topology shown in Figure 4–1(b) will operate under several protocols, but the only types to be considered in the following chapters are the CSMA/CD and token-passing types. Ring topologies lend themselves well to baseband TDM transmission techniques. The nodes on the ring serve as repeater stations, which also have error-checking circuitry. The delay at these stations can be as little as 1 bit. The upper limit is dependent on buffers allocated at each node.

The third topology is a star, a technique that has been used by a number of computer manufacturers for many years. For example, when several terminals are connected to a computer, a star is the easiest method for implementation.

Tree topologies can be composed of bus systems similar to Wangnet for the broadband case and Ethernet for a baseband application. But one should always beware of strict categorization of networks, because networks can be designed as combinations of rings, stars, buses, and trees.

One example of a network composed of two rings, which may be viewed as a star, was constructed by the author [see Figure 4–2 (a)]. When the transmission delay between nodes is small compared to bit time, the implements will appear to be connected together. For optical waveguides, the propagation delay is about 5 nanoseconds/meter (ns/m). If a 1-Mbit/s transmission rate is used on the fiber-optic ring, then 10- to 20-meter distances will have a 5 to 10 percent bit time delay. See Equation 4–1.

$$\tau = \frac{N_{core}}{C} = \frac{1.50}{3 \times 10^8}, \text{ propagation delay in glass} \qquad (4\text{--}1)$$

$$\tau = 5 \text{ ns/m}$$

If the copper ring is also kept small, then the network can again be assumed to be the node of a star. The result of the reduction is shown in Figure 4–2(b). The branch connecting N_1 to N_2 is part of the gateway station. The objective of this discussion is to make the reader aware that network architectures may allow variations in topology (i.e., more than one topology may be suitable for the analysis).

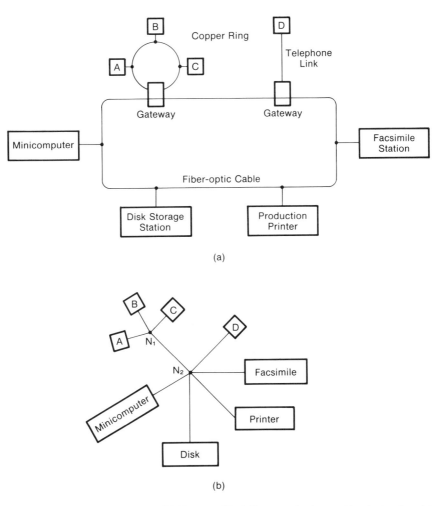

Figure 4–2 (a) Loop cable plant. (b) Topographical diagram of a loop cable plant reduced to two stars.

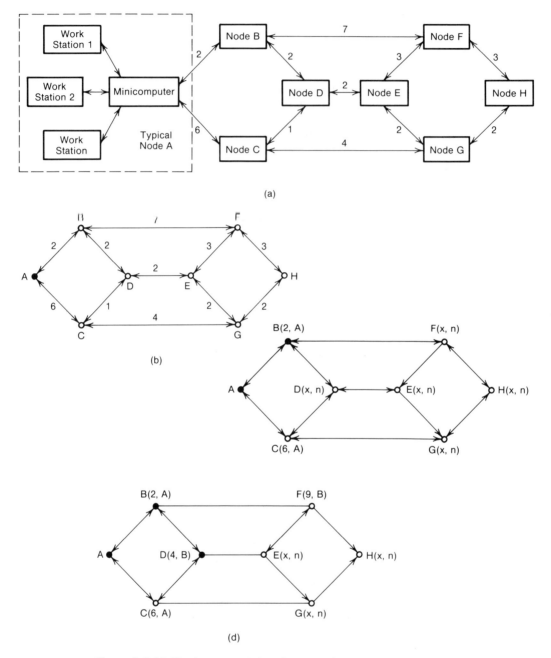

Figure 4–3 (a) Simple representation of a network with distances between the nodes indicated. (b) Graphical representation of Figure 4–3(a). (c) Second permanent node selected. (d) Third permanent node selected. (e) Fourth permanent node selected. (f) Final graph with all the permanent nodes shown with minimum path reduction.

Graph Theory

In this section a discussion will be presented to give the reader an appreciation of what can be accomplished on a small-scale network. Larger networks require a computer to perform an analysis, but a study of smaller networks will allow the reader to make some intuitive assessment of these networks, at least. The analysis will examine flow of data. Flow analysis may be found in other disciplines, such as electrical engineering (network synthesis and analysis), fluid mechanics, servosystems, and mathematics, to name a few.

Let us first define the terms to be used in the analysis. A node is the termination of a branch (such as a star with a branch radiating from the center terminated in a workstation) or an intersection of multiple branches [such as a star; see Figure 4–1(c)]. For ring topologies, nodes are the workstations located on the ring, while in bus networks the nodes are taps along the cable. Branches are the line segments connecting nodes, which may be unidirectional or bidirectional, with the former considered a directed branch and the latter an undirected branch. Networks may be composed of only directed branches, a simplex ring, where the flow of data is in one direction. The other case would be a star, with communication in both directions. A mixture of both types of branches would be a computer connected in a star with terminals and RO (receive only) printer stations.

Let us examine the network in Figure 4–3(a) and determine the shortest path

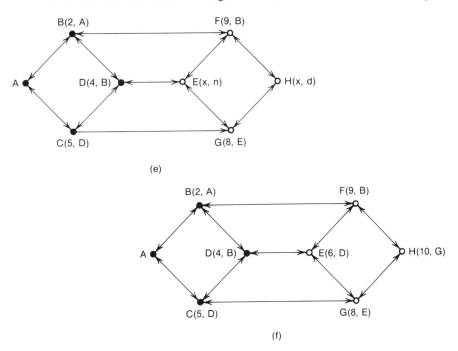

Figure 4–3 (Continued)

between two nodes within it. Figure 4.3(a) is a physical representation of a network, with Figure 4.3(b) as the graphical equivalent. The graph will be used to find the shortest path from A to H. The node that is the starting point is filled in as shown in Figure 4-3(b). The adjacent nodes are next labeled with working node (A) and the distance from it, as shown in Figure 4-3(c). The nodes are then searched for the smallest distance between the working node and all adjacent nodes, which for the case in Figure 4-3(c) is node B. Node B is then considered permanent, as shown in Figure 4-3(c), with node B filled in. Note, that C is not, with all other nodes unknown (x,n). The third permanent node is selected as D because the distance from A to D is shorter than A to C or A to F. If the graph is examined for the next permanent node, the reader will observe that the distance A to C via node D is the shortest distance from A as compared to all other tentative nodes. Hence it is labeled as a permanent node [see Figure 4-3(e)]. The completed graph is shown in Figure 4-3(f).

Some observations should be made. There are occasionally multiple shortest paths, such as in the case of node F. Paths ABF and ABDEF are both nine units long, although the former only passes through three nodes, which should be the choice because the nodes often present additional delays in the network. Therefore, a delay will make the transmission path appear to be longer. A procedure for computing the shortest path is shown in Figure 4-4, which is an algorithm in Pascal. See reference [1] for further study of this subject. This simplified analysis assumes that all transmission rates are identical, which may not always be the case.

If all transmission were to take the shortest path, a particular branch may become overloaded, thus causing excessive delays. The next topic will address network data flow in bits per second, which is a traffic analysis of the network.

Network Flow

To compute the information-carrying capacity of a network, the cut will be used. This is a technique of removing (cutting) branches between two nodes until they are disconnected.

Let us examine the cuts shown in Figure 4-5(a).

Cut 1	AB, AE	11 Flow units (fu)
Cut 2	AB, ED, JF, JK	23 Flow units (fu)
Cut 3	BC, FG, KL	10 Flow units (fu)
Cut 4	CH, LH	12 Flow units (fu)

These are not the only cuts that could have been made (for example, AB, ED, EI, BC, GC, and LH, among others.) The maximum possible flow across the network is 10 fu because the flow units in each branch are considered the maximum capacity of that branch. Before we proceed any further, a glaring problem should immediately be apparent. The maximum flow into B is 9 fu and the flow

```
const n = ... ;                          {number of nodes}
      infinity = ... ;                   {a number larger than any possible path length}
type node = 0 .. n;
     nodelist = array [1 .. n] of node;
     matrix = array [1 .. n, 1 .. n] of integer;

procedure ShortestPath (a: matrix; s,t: node; var path: nodelist);
{Find the shortest path from s to t in the matrix a, and return it in path.}
type lab = (perm, tent);                 {is label tentative or permanent?}
     NodeLabel = record predecessor: node; length: integer; labl: lab end;
     GraphState = array [1 .. n] of NodeLabel;
var state: GraphState;   i,k: node;   min: integer;
begin                                    {initialize}
  for i := 1 to n do
    with state [i] do
      begin predecessor := 0; length := infinity; labl := tent end;
  state [t].length := 0;   state [t].labl := perm;
  k := t;                                {k is the initial working node}
  repeat   {is there a better path from k?}
    for i := 1 to n do
      if (a[k,i] <> 0) and (state [i].labl = tent) then    {i is adjacent & tent.}
        if state [k].length + a[k,i] < state [i].length then
          begin
            state [i].predecessor := k;
            state [i].length := state [k].length + a[k,i]
          end;
    {Find the tentatively labeled node with the smallest label.}
    min := infinity;   k := 0;
    for i := 1 to n do
      if (state [i].labl = tent) and (state [i].length < min) then
        begin
          min := state [i].length;
          k := i                         {unless superseded, k will be next working node}
        end;
    state [k].labl := perm
  until k = s;                           {repeat until we reach the source}
  {Copy the path into the output array.}
  k := s;   i := 0;
  repeat
    i := i + 1;
    path [i] := k;
    k := state [k].predecessor;
  until k = 0
end;   {ShortestPath}
```

Figure 4–4 Pascal program for finding shortest path between the source and sink of a network.

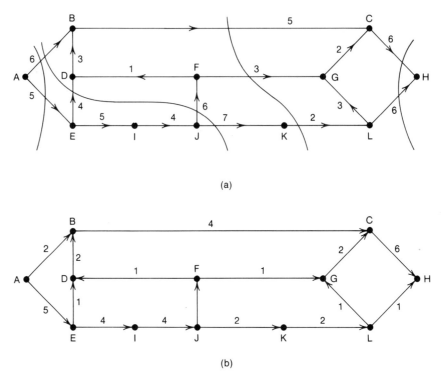

Figure 4–5 (a) Unfeasible network showing cuts. (b) Feasible network showing cuts with no rules violated.

out is 5 fu, which implies the node is a sink; this is not the case for our analysis. Also note that node E has 5 fu in and 9 fu out, which implies it is a source, which is not the case for analysis given here. Therefore, some general rules are formulated to aid in the analysis.

1. The source has no inward branches.
2. The sink has no outward branches.
3. No branch has more flow than its capacity, but may have less.
4. The inflow is equal to the outflow except for sources and sinks.
5. The outflow of a source must equal sink inflow.

The conditions given are necessary for the network to be feasible. Let us now examine the network in Figure 4–5(b) to see how the network can be made feasible. Note that all the conditions are met in Figure 4–5(b). Examining the original cuts shown in Figure 4–5(a), we see that they all have the same flow of 7 fu in Figure 4–5(b). Another rule to be added to the analysis was already mentioned previously in a less formal manner.

6. The maximum flow between any two arbitrary nodes in any graph cannot exceed the capacity of the minimum cut separating these two nodes. [Minimum cut is a cut with minimum flow across it, such as cut 3 in Figure 4–5(a)].

A more formal methodical approach is needed for finding the maximum flow and minimum cut. A maximum flow algorithm was published in 1978 by Mathotra, et al.

A Pascal program for calculating maximum flow is shown in Figure 4–6. A simplified analysis showing the flow of the algorithm is given in reference [1], but the model only assumes a single source–sink pair. The model is adequate for explaining some of the algorithm's subtleties, but in a realistic world one must deal with large numbers of source–sink pairs. Large-scale analyses of this type are beyond the scope of this text.

Node-failure analysis is perhaps one of the most important topics of this book. The objective is to remove paths from the network and examine the reliability issues. Network failures can occur due to any number of reasons, such as normal electronic failures, software glitches in programs, destruction of a facility due to natural disaster, or destruction of nodes in military situations (the tactical case). Degrees of reliability are determined by the end user and must be considered on a per network basis.

In some instances node failure presents no immediate problem, such as the case for a sales office. But for tactical situations, hospital patient monitors, and banking, extremely high reliability is required. Often, in military situations, sufficient redundancy is employed to prevent catastrophic failure, but the networks operate at a reduced performance. Reliability can be increased manyfold by adding 100 percent redundancy, but the cost becomes manyfold also. Most installations use the reduced-performance technique, which allows peak performance the majority of the time.

If two nodes are to be disconnected, the minimum number of branches removed to disconnect them is their minimal cut set.

Two paths are considered to be disjoint (i.e., branch disjoint) if they do not share a common branch. Branches may share a common node and be considered branch disjoint. If Figure 4–7(a) is examined carefully, one will note paths ADF and ABDCEF are branch disjoint, but paths ABDCEF and ADCGF are not because of the DC segment. The graph represented by Figure 4–7(b) is a four-path branch-disjoint network connecting node X to Y. If at least one branch is removed from each path, X and Y will be disconnected; then, intuitively, we would expect for K branch-disjoint paths that at least K branches must be removed to disconnect X and Y. Identifying the number of branch-disjoint paths expresses a bound on the minimum cut.

The maximum-flow algorithm may be used to calculate the number of branch-disjoint paths. Replace the weight of each branch with the weight of 1. When a calculation is made to find the maximum flow between the source and sink, the resulting flow will be equivalent to the minimum cut, which is also equal to the number of branch-disjoint paths.

```
const n = ... ;
    unscanned = −n;
    infinity = ... ;
type node = 1 .. n;
    xnode = −n .. n;
    vector = array[node] of xnode;
    matrix = array[node, node] of real;
    WhichWay = (push, pull);

procedure MaxFlow(s,t: node; c: matrix; var f: matrix);
var RefNode: node;              {node with least excess capacity}
    MinPotential: real;         {excess capacity of the reference node}
    layer: vector;              {the layered network is defined by this array}
    i,j: node;                  {indices}

function min(x,y: real):real; begin if x < y then min := x else min := y end;

procedure walk(i: node);
{Traverse the layered network from t, inverting layer numbers.}
var j: node; li: xnode;
begin
    layer[i] := −layer[i];
    li := layer[i];
    for j := 1 to n do
        if (j <> s) and (−layer[j] = li − 1) and ((f[j,i] < c[j,i]) or (f[i,j] > 0))
            then walk(j)
end; {walk}

function LayeringPossible: boolean;
{Is it possible to build a layered network? If so, build it.}
var i,j: node;  k: 0 .. n;  EmptyLayer: boolean;
begin k := 0;                       {k keeps track of layer being built}
    for i := 1 to n do layer[i] := unscanned;  {initialize each node}
    layer[s] := k;                  {source node is in layer 0}

    repeat
        k := k + 1;                 {now locate all nodes in layer k}
        EmptyLayer := true;         {an empty layer stops the algorithm}
        for i := 1 to n do
            if −layer[i] = k − 1 then
                {i is in layer k − 1, its neighbors may be in layer k.}
                for j := 1 to n do          {check each node adjacent to i}
                    if (layer[j] = unscanned) and ((f[i,j] < c[i,j]) or (f[j,i] > 0))
                        then begin layer[j] := −k; EmptyLayer := false end;
    until (layer[t] <> unscanned) or EmptyLayer;
    LayeringPossible := not EmptyLayer;
    walk(t)                         {prune off the dead ends}
end; {LayeringPossible}
```

Figure 4–6 Pascal representation to compute maximum flow in a network.

procedure *FindRefNode*(*i*: *node*);
{Traverse the layered network from *i*, seeking the reference node.}
var *j*: *node*; *li,lj*: *xnode*; *InCap,OutCap*: *real*;
begin
  li := *layer*[*i*]; *InCap* := 0; *OutCap* := 0;
  for *j* := 1 **to** *n* **do**   {examine each node adjacent to *i*}
    begin *lj* := *layer*[*j*];
      if (*lj* = *li* − 1) **and** (*j* <> *s*) **and** ((*f*[*j,i*] < *c*[*j,i*]) **or** (*f*[*i,j*] > 0))
        then *FindRefNode*(*j*);
      if *lj* = *li* − 1 **then** *InCap* := *InCap* + (*c*[*j,i*] − *f*[*j,i*]) + *f*[*i,j*];
      if *lj* = *li* + 1 **then** *OutCap* := *OutCap* + (*c*[*i,j*] − *f*[*i,j*]) + *f*[*j,i*]
    end;
  if (*i* <> *s*) **and** (*i* <> *t*) **and** (*min*(*InCap,OutCap*) < *MinPotential*) **then**
    {Node *i* has a smaller potential than the current reference node.}
    begin *MinPotential* := *min*(*InCap,OutCap*); *RefNode* := *i* **end**
end; {*FindRefNode*}

procedure *PushPull*(*i*: *node*; *FlowLeft*: *real*; *p*: *WhichWay*);
{Augment the flow through *i* by pushing or pulling *MinPotential* units.}
var *j,k1,k2,LayerSought*: 0 .. *n*; *r*: *real*;
begin *j* := 0;
  while (*FlowLeft* > 0) **and** (*j* < *n*) **do**
    begin *j* := *j* + 1;
      if *p* = *push*
        then begin *k1* := *i*; *k2* := *j*; *LayerSought* := *layer*[*i*] + 1 **end**
        else begin *k1* := *j*; *k2* := *i*; *LayerSought* := *layer*[*i*] − 1 **end**;
      r := *min*(*FlowLeft*, *c*[*k1,k2*] − *f*[*k1,k2*] + *f*[*k2,k1*]); {am't of flow to move}
      if (*r* > 0) **and** (*layer*[*j*] = *LayerSought*) **then**
        begin   {push/pull some flow to/from an adjacent layer}
          FlowLeft := *FlowLeft* − *r*;
          f[*k1,k2*] := *f*[*k1,k2*] + *r* − *min*(*r,f*[*k2,k1*]);   {augment positive flow}
          f[*k2,k1*] := *f*[*k2,k1*] − *min*(*r, f*[*k2,k1*]);   {push reverse flow bkwrd}
          if (*j* <> *s*) **and** (*j* <> *t*) **then** *PushPull*(*j,r,p*)
        end
    end
end; {*PushPull*}

begin
  for *i* := 1 **to** *n* **do for** *j* := 1 **to** *n* **do** *f*[*i,j*] := 0;   {initially no flow}
  f[*s,t*] := *c*[*s,t*];   {if an *s-t* link exists, saturate it}
  while *LayeringPossible* **do**   {assign nodes to layers}
    begin *MinPotential* := *infinity*;
      FindRefNode(*t*);   {find the reference node}
      PushPull(*RefNode*, *MinPotential*, *push*);   {push flow toward *t*}
      PushPull(*RefNode*, *MinPotential*, *pull*);   {pull flow from *s*}
    end
end; {*MaxFlow*}

                                                                maximum flow.

Figure 4–6 (Continued)

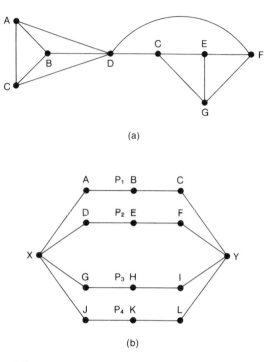

Figure 4-7 (a) A network flow graph. (b) A flow graph of a completely branch-disjointed network.

The discussion has been directed to branches being removed to determine graph connectivity. If nodes are removed, eventually the source and sink will be disconnected, and this will determine the node connectivity of the graph.

The maximum-flow algorithm may also be used to calculate the node-disjoint paths between the source and sink. The original undirected graph is modified into a directed graph with $2n$ nodes (where n = nodes of the original graph) and $2b + n$ branches (b = original graph branches). Each node is replaced by two nodes labeled n and n^1. All incoming directed branches are connected to n, and all outgoing branches are attached to n^1. An example of a node-disjoint path is in order to show a simplified case.

Figure 4-8(a) is the original network with bidirectional branches connecting all nodes. If each node is expanded into n and n^1, the network will resemble Figure 4-8(b). From Figure 4-8(a), it can be noted that the node count is 8 and 16, or $2n$ for Figure 4-8(b). The number of branches is 11 in Figure 4-8(a) and 30 ($2b + n$) for Figure 4-8(b), which is correct.

The maximum-flow program will find the node-disjoint paths between the source and sink (X and Y, respectively). Each branch will have a weight of 1 in the analysis. The node-disjoint paths in Figure 4–8(b) are $X\bar{X}B\bar{B}E\bar{E}Y$ and $X\bar{X}A\bar{A}F\bar{F}Y$. The key to the analysis is that each unit branch can have only one path through it; therefore, no original node may be on two or more paths. The maximum-flow algorithm will determine all the paths through the network that use sets of nodes not used by other paths (i.e., each path has a unique set of nodes). If this algorithm were run for each set of nodes to determine the node-disjoint paths for a large network (e.g., a 20-node network), it would require 190 computer runs. Therefore, this technique is useful for finding node-disjoint paths for only 5 or 6 sets of nodes.

In a real situation such as encountered in tactical networks, nodes and branches can be removed or destroyed and the network connectivity must be

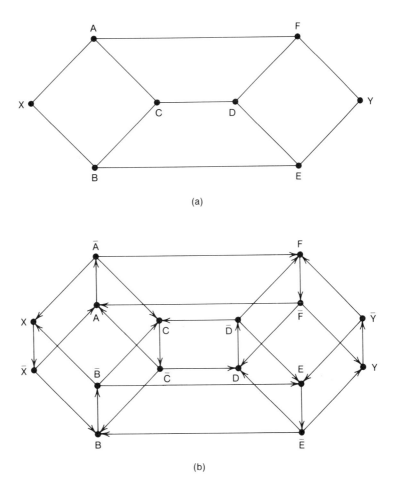

Figure 4–8 (a) Original network with all branches bidirectional. (b) Network 4–8(a) reconstructed with unidirectional branches.

```
program ReliabiltySimulation(input, output);
const n = ... ;
type matrix = array[1 .. n, 1 .. n] of boolean;
var max,nrounds,np,ip,discon,i,j,k,nr: integer;
    p, NumDiscon, FracDiscon, pincr, seed: real;
    a,b: matrix;

function random: real;              {random number from 0.0 to 1.0}
begin
  seed := 125.0 * (seed + 1.0);
  seed := seed − 8192.0 * trunc(seed/8192);
  random := (seed + 0.5)/8192
end; {random}

begin
  max := n*(n − 1) div 2;           {number of possible communicating pairs}
  read(nrounds, p, pincr, np, seed);  {input data}
  for i := 1 to n do for j := 1 to n do
    begin read(k); if k = 0 then a[i,j]:=false else a[i,j]:=true end;

  for ip := 1 to np do              {loop, increasing p each time}
  begin
    NumDiscon := 0;                 {number of disconnected networks}
    FracDiscon := 0;                {fraction of disconnected pairs}
    for nr := 1 to nrounds do       {make multiple simulations per value of p}
      begin
        b := a;                     {copy the entire array a to b}
        for i := 1 to n do          {destroy links at random}
          for j := i + 1 to n do
            if b[i,j] then
              if random < p then begin b[i,j] := false; b[j,i] := false end;

        for i := 1 to n do          {compute connectivity of the graph}
          for j := 1 to n do
            if b[j,i] then for k := 1 to n do b[j,k] := b[j,k] or b[i,k];

        {Check to see how many pairs are still connected.}
        discon := 0;                {count disconnected pairs}
        for i := 1 to n do
          for j := i + 1 to n do
            if not b[i,j] then discon := discon + 1;

        FracDiscon := FracDiscon + discon/max;
        if discon > 0 then NumDiscon := NumDiscon + 1.0
      end;

    writeln(p: 6:2, FracDiscon/nrounds: 6:2, NumDiscon/nrounds: 6:2);
    p := p + pincr                  {increase p to next value}
  end
end.
```

Figure 4-9 Pascal program for computing network reliability by simulation.

known. Also, it must be known immediately when the destruction occurs. The calculation must be made well in advance, because in an emergency situation computers may not be available. With the advent of low-cost microprocessors and electrically erasable read-only memories (E^2PROMs), connectivity tables may be stored at the nodes. This will be discussed when controllers are addressed in Chapter 6.

A particularly useful program is shown in Figure 4–9. A Pascal program is given due to its popularity as a programming language. A comprehensive explanation of the program's use may be found in reference [2]. The simulation includes both node and branch failures, and it is based on the Monte Carlo method for selecting random failures.

During the simulation, additional branches may be added to reduce the probability of connectivity failures. If additional nodes and branches are added to strategic points in the network, which will increase the traffic throughput, the network will become more efficient and reliable. Cost, reliability, performance, and efficiency are all prime movers in the design of many systems. Therefore, the designer must make choices dictated by the design requirements. Any increases in the number of branches or nodes will be accompanied by an increase in cost. A cost–performance ratio will generally indicate to the designer when an optimum is being approached. When increases in network nodes or branches cause only small changes in reliability and performance, but result in large cost increases (the approach of diminishing returns), the network is near optimization.

A variation of the simulation method can be stored in E^2PROM with a connectivity table. After any new branches or nodes are added to the network, the simulation can be executed and the connectivity table updated. The program may actually be on disk, with the table only stored in E^2PROM. This would allow each node to upgrade its own connectivity graph.

Another variation for upgrading connectivity tables can be accomplished using telephone modems. They can connect with the node terminals during low traffic hours (e.g., on weekends). The conductivity simulation can be run and the upgrades programmed into the nodes. This technique could be made to run automatically. One fast-food chain uses a similar technique to program their cash registers during the night. This upgrading technique is transparent to the node terminal operators unless an actual emergency arises.

The previous text addressed connection availability and to some extent performance, but in a superficial context. The flow that can be related to bandwidth and message or data transmission time must be examined. Also, an important performance criterion is delay, which is to be addressed in detail in Chapter 6.

A study of traffic density is also necessary to further enlighten the reader as to how performance is affected by traffic conditions. As may be observed from some of the previous results, many nodes may use the same path, which is the optimum, and the traffic density may be so great that the network may not function properly, if at all. Therefore, prior to any physical design, the ramifications of these conditions must be known to produce a functionally well designed network.

Time-delay Analysis

The network must now be examined for the time-delay impact. Delay in networks can be subdivided into three major categories:

1. Delay due to the physical attributes of the network, such as propagation delay, delay specifically induced by logic circuits (digital), or delay lines and filters (analog delay).
2. Delay due to software, which may be inherent in a program or added for a specific purpose.
3. Delay attributed to operator interoperability delays, for example, due to operator typing answers to a query while on line. This adds human reaction time to delay, which is in general considerably longer than the other forms.

The first two types of delays are covered in detailed form in the text, while the third is superficially addressed. Human engineering issues are more application dependent than the first two types of delay. Texts on human engineering are references [3], [4], and [5].

Let us begin our analysis with the time delay caused by propagation of a signal down a fiber-optic or copper transmission line. Propagation in a fiber-optic cable is given in Equation 4–2.

$$\tau = \frac{n}{c} \text{ s/m} \qquad (4\text{–}2)$$

where n = refractive index
$c = 3 \times 10^8$ m/s, speed of light in a vacuum
Thus

$$\tau = \frac{1.5}{3} \times 10^{-8} \text{ s/m or 5 ns/m}$$
$$\text{or}$$
$$\tau = 5 \text{ns/km}$$

The equation is only referring to delay due to the transmission line and not to pulse spreading, which will be discussed later. The assumption here is that the transmission medium is a uniform glass rod.

In copper lines the velocity constraint must be considered when calculating transmission-line delay. When electromagnetic waves travel down a transmission line that uses spacing insulators or solid dielectrics, transmission decreases in velocity as compared to in a vacuum. The waves are affected by the inductance and capacitance of the transmission line. Some typical velocity constants are as follows:

1. Parallel lines with an air dielectric between them: $0.95c$ to $0.97c$
2. Parallel lines with a plastic dielectric between them: $0.80c$ to $0.95c$
3. Shielded pair with rubber insulation: $0.56c$ to $0.65c$
4. Coaxial line with air dielectric: $0.85c$
5. Coaxial line with plastic dielectric similar to CATV: $0.77c$
6. Twisted pair, rubber insulation: $0.56c$ to $0.65c$

For a coaxial line, the velocity of the electromagnetic wave is similar to that of the glass waveguide:

$$v = 0.77 \times 3 \times 10^8 \text{ m/s}$$
$$= 2.21 \times 10^8 \text{ m/s}$$
$$\tau = 4.1 \text{ ns/m}$$

An interesting note: water has an index of refraction of 1.33 or a propagation time of $\tau = 4.4$ ns/m, which is very close to that of good-quality coaxial cable.

As one may observe, this is the one delay parameter that cannot be easily controlled as compared to the other two types. Once a cable plant is installed, the cost of replacing it with shorter-propagation-time types of cable would be cost prohibitive. The delay, as will be demonstrated in Chapter 6, will have an impact on frame size and LAN efficiency.

As a simple example of propagation effect, let us consider a LAN with a transmission rate of 10 Mbit/s and further assume that the data are sent in 500-bit packets. The packet represents a line length of 10 km for the fiber-optic case and 12 km for coaxial cable. This implies that two stations communicating in half duplex must listen at least 50 microseconds (µs) before transmitting, if the distance between them is 10 to 12 km and no other delays are present. If an acknowledge is required prior to transmission of another packet, the transmitting station will require a 100-µs delay before transmitting. This latter case is not efficient, and there are methods of circumventing this acknowledge delay. Also, as packet length increases, the propagation delay begins to become less significant. Therefore, it is perhaps intuitive that, as transmission distances become greater, packet size should also be made larger. When designing a LAN, the engineer must allow for physical expansion or he or she may soon have a very inefficient network if cable plants are expanded. Making packets larger after a network is complete can be very cost prohibitive and cause a multitude of other problems.

The next type to be considered is delay as data pass through a node due to the hardware (digital logic). This is also physical delay. There are several reasons for this, as follows:

1. The removal of extra bits in the data stream due to bit stuffing. Bit stuffing is a method of assuring that transitions occur if long strings of zeros or ones occur in data. This will be examined on a specific system in Chapter 6.
2. Delays due to cyclic redundancy checking (CRC).
3. Delays due to clocking of incoming data to the transmitter in the case of a repeater.
4. Decoding and encoding delays.

As an example of the effect of these delays, let us consider a bit-stuffing situation. Each station on a LAN has a 5-bit delay due to bit stuffing (i.e., no more than five consecutive ones can occur in the data). The bit stuffer will insert a zero to force a transition in the data stream if more than five consecutive ones occur. Long strings of consecutive zeros have no effect because zeros have transitions. If the data pass through 20 nodes and each has this 5-bit delay and the data transmission rate is 10 Mbits/s, the delay due to the nodes will be 10 μs. Note that the logic delay is insignificant compared to propagation delay, but if the transmission rate is reduced to 1 Mbit/s, the reverse is true (i.e., the propagation time is insignificant and the node delay becomes an appreciable 100 μs). This part of the physical delay can be controlled by the designer more easily than cable delays.

The next type of delay is attributed to software. An example of software delay is when messages are stored in a buffer and are assembled for transmission. The packets may be prepared and held in the buffer until the total message is assembled before it is transmitted or, if the node is acting as a repeater, the entire message may be received before it is retransmitted. For this latter situation, the calculations are shown in Example 4–1.

Example 4–1

Let the message equal 1 page of text.

$$\text{Page of text} = 78 \text{ characters} \times 48 \text{ lines}$$
$$\approx 4000 \text{ characters with overhead}$$
$$\text{Characters in ASCII} = 4000 \text{ bytes}$$
$$\text{Number of bits} = 8 \times 4000$$
$$= 32{,}000 \text{ bits}$$

For a 10-Mbit/s transmission rate, the delay is 3.2 milliseconds (ms).

From the example, it is obvious that this type of software delay makes all the physical delays insignificant.

The final type of delay, that due to the human engineering factors, is even larger. Let us examine a typical example.

Example 4–2

A typist can generally type at 70 to 120 words/minute. If this speed is encoded in ASCII, the typing rate is TY_R.

$$TY_R = 6 \text{ characters/word} \times 100 \text{ words/minute} \times \frac{\text{minute}}{60 \text{ s}} \times \frac{8 \text{ bits}}{\text{char.}}$$

$$TY_R = 80 \text{ bits/s}$$

If an acknowledgment were to be typed after each message and it only required 6 characters, the 0.6-s response would be much larger than any of the previous delays.

Another type of delay found in networks, which may be implemented in hardware, software, or a combination of both, is queuing. The queue is transparent to the human operator at the node. It is a buffer such as shown in Figure 4–10(a) and (b). The queue will allow the node to use the transmission facilities more efficiently. As an example, if only one message were handled at the node at a time, the transmission would have bursts of high-speed data and long idle channel periods. In Figure 4–10(a), messages may arrive randomly, be stored at the input queue, and then be sorted by the controller for transmission according to time of arrival or priority. Figure 4–10(b) shows a queue where the arrival data may be immediately sent to the output queue, or they may be held until the node finishes transmission before being repeated.

Messages are composed of packets, which can be described by the following equation:

$$m_i = \sum_{n=1}^{k} P_n \qquad m_i \text{ may be composed of a single packet}$$

Let us now generate the equation that governs queues. The message arrival rate is the sum of the individual arrival rates, as shown in Equation 4–3.

$$M = \sum_{i=1}^{k} m_i \quad \text{messages/second} \qquad (4\text{–}3)$$

Then, if l = average bits per message,

$$T_r = M_l \quad \text{bits/second} \qquad (4\text{–}4)$$

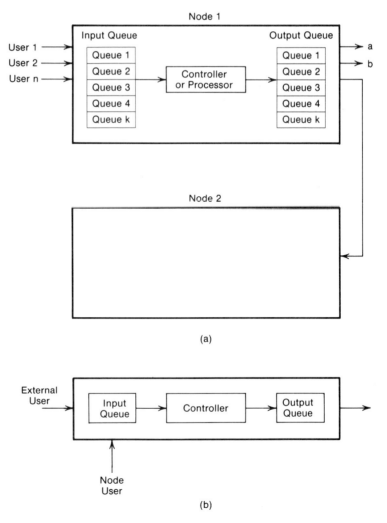

Figure 4–10 Queues for (a) star and (b) ring connections.

Equation 4–4 is the transmission rate of the traffic, which is usually not operating at full capacity, C_m, the maximum transmission rate of the channel.

$$U = \frac{T_r}{C_m} \times 100 = \frac{M_l}{C_m} \times 100, \quad \text{utilization of the channel (\%)} \quad (4\text{--}5)$$

The service time τ_s required for the node server (controller) to process a message is given by Equation 4–6.

$$\tau_s = \frac{l}{C_m} \text{ seconds, average service time} \qquad (4\text{–}6)$$

Also, $U = M\tau_s$ from Equation 4–5.

Let us determine the average waiting time τ_w of a message in the queue. If N is the average number of messages in the queue, then

$$\tau_w = N\tau_s, \quad \text{waiting time} \qquad (4\text{–}7)$$

Equation 4–7 is intuitive. Since it takes τ_s to service a message, then if N messages are in the queue it takes $N\tau_s$ time to service all of them before the next entering message can be serviced.

If the message rate entering the input queue is equal to the message rate leaving, the queue will remain the same length. If, however, the message rate at the input should vary (i.e., increase or decrease) and the output rate remains the same, the queue length will change. The probability of new message arrivals must be known if a realistic queue waiting time is to be calculated in a dynamic situation.

If the message arrivals are at k intervals and a Poisson's distribution is used to provide the probability of the number of new arrivals, this can be expressed by Equation 4–8.

$$P(k) = \frac{\lambda^k e^{-\lambda}}{k!} \quad k = \text{arrivals in } \tau_s \text{ seconds} \qquad (4\text{–}8)$$

$$U^1 = \tau_s M = \frac{M_l}{C_m} U, \quad \text{not in \%}$$

where $\lambda = \tau_s M = U^1$. Then

$$P(k) = \frac{(U^1)^k e^{-U^1}}{k!} \qquad (4\text{–}9)$$

The version of Poisson's distribution depicted in Equation 4–9 is dependent on channel utilization (more meaningful), provided that the service time remains fixed, which is the usual situation for LANs. For networks such as telephone systems, the service time may vary because multiple servers are at each node (telephone exchanges).

Let us now examine the queue based on a steady-state operation (i.e., it is not growing or shrinking in size). The number of messages in the queue can be estimated based on Equation 4–10.

$$N = \frac{U^1}{\Delta U^1}, \quad \text{where } \Delta U^1 = 1 - U^1 \qquad (4\text{--}10)$$

$$\tau_w = N\tau_s = \frac{U^1}{\Delta U^1}\tau_s$$

$$\tau_w = \frac{l}{C_m}\left(\frac{U^1}{1 - U^1}\right) \qquad (4\text{--}11)$$

Equation 4–11 is valid provided the value of N does not become so large that the node lacks adequate buffer space. It is interesting to note that as U^1 approaches 1, which implies 100 percent utilization of the channel capacity, the queue N will approach infinity; this is obviously not the case. Therefore, 100 percent utilization of the channel is not possible in Equation 4–11. But when various types of protocols are discussed, we will discover means for circumventing this problem (e.g., if the queue is full, the network will busy out the new arrivals).

The total average delay time attributed to the queue is given by Equation 4–12.

$$T_t = \tau_s + \tau_w$$

$$= \tau_s(1 - N) = \tau_s\left(1 - \frac{U^1}{1 - U^1}\right) \qquad (4\text{--}12)$$

$$= \frac{\tau_s}{1 - U^1}$$

$$= \frac{l}{C_m}\left(\frac{1}{1 - U^1}\right) \qquad (4\text{--}13)$$

$$= \frac{l}{C_m}\left[\frac{1}{1 - M(l/C_m)}\right] \qquad (4\text{--}14)$$

$$= \frac{l}{C_m - M_l} \quad \text{(alternative)} \qquad (4\text{--}15)$$

The alternative equation is given to allow the reader to observe some of the trade-offs, for example, message length and total average queue delay time. If the derivative of Equation 4–15 is taken with respect to l and the minimum is calculated to optimize M, the equation is 4–16

$$\frac{dT_t}{dl} = 0 = \frac{(C_m - M_l) + M}{C_m - M_l}$$

$$M_{opt} = \frac{C_m}{l - 1}, \quad \text{optimized value for } M \qquad (4\text{--}16)$$

Channel Capacity

Channel capacity is another parameter that must be examined. The reason for this concern is that the designer must allow sufficient capacity to expand the facilities, whereas overdesign increases the cost of plant facilities, which in turn results in inefficient use. The maximum capacity of a transmission system can be calculated using the equation stated in Shannon's paper (see reference [3]).

$$C_m = \text{BW} \log_2 (1 + \text{SNR}) \qquad (4\text{--}17)$$

where BW is bandwidth and SNR is the signal-to-noise ratio.

Shannon stated in his paper that, for a band-limited channel with $C_i \leq C_m$, a code exists for which the error rate approaches zero as the message length approaches infinity. Also, conversely, if $C_i > C_m$, the error rate cannot be reduced below some positive numerical limit.

Example 4–3

$$C_m = 3000 \log_2 (1 + 1023)$$
$$= 30{,}000 \text{ bits/sec}$$

Applying Equation 4–17 to a typical voice channel with an SNR of 1023 and 3-kHz bandwidth, this value is the absolute maximum, but it cannot be achieved because of intersymbol interference. This equation does put an upper bound on the system, however, which can be useful to rule out impossible values.

An improved equation for calculating C_m with intersymbol interference accounted for is Equation 4–18.

$$C_m = 2\text{BW} \log_2 M \text{ bits/s} \qquad (4\text{--}18)$$

where M = number of levels. This equation is attributed to Nyquist.

The equation does not consider bit error rates (BER). As the number of levels M increase, larger SNR is needed to resolve the signals. The SNR-to-BER relationship will be covered in Chapter 5. As an application example using Equation 4–18, let $M = 16$ and BW = 3 kHz; then C_m = 24 Kbits/s.

When a 16-level system with phase-shift keying is used and the transmission rate is 2400 bits/s, it is also possible to convey information at a 9600-bit/s rate. This is accomplished in many present-day audio modems. However, phase coding does not come without a penalty. Each time the phase information is doubled, a 3-decibel (dB) increase in noise penalty results.

Some of the more sophisticated modems are available with higher transmission rates (19.2 kHz bits). These modems do not operate over standard switched networks, but require conditioned lines; these are leased from telephone companies at a premium cost. The conditioning guarantees minimum attenuation and phase distortion. The conditioning is accomplished through predistortion of the signals, commonly known as equalization. Some newer modems incorporate equalizers and the line prior to transmission and add equalization as necessary. This is adaptive equalization. Some of these newer modems test the line after it is equalized. If the BER is below a threshold value, the modem shifts down in transmission rate until the BER is adequate. The procedure takes several seconds, but the process allows the modem user to take full advantage of exceptionally good lines. Quite often in business, data transmission calls may be for short distances, say within a campus. Under these conditions, communications are extremely good as compared to an East Coast to West Coast connection. The cost of these adaptive modems is generally 15 to 20 times greater than manually selected transmission rate modems with no equalization.

Adaptive modems are designed with microprocessors and hosts of interface circuits. The author expects the cost of these modems to drop significantly as gate arrays and analog circuit arrays replace microprocessor interface circuitry. A design will be presented in Chapter 5 under integrated-circuit controllers.

The previous discussion mainly dealt with telephone lines, but many LANs do not use these facilities. Most LANs have a gateway station, which is useful for communicating with other LANs via telephone lines; therefore, for these situations the modem arguments hold. But the majority of LANs require gateway stations that will allow communication between two different types of LANs (e.g., Ethernet and ARPANET). The capacity equations 4–17 and 4–18 hold, but modems are quite different from the telephone type. For example, these modems require no line equalization, and the specification of cable plant facilities is much more stringent (i.e., delay and amplitude distortion have much closer tolerances).

Noise Considerations

Noise contributions from various devices in a network will degrade transmission facilities. The devices are amplifiers, receivers, cable plants themselves, filters, multiplexers, and many others. Our discussion of noise will address the normal types of noise found in copper cable systems and then optical noise components.

Thermal noise occurs in all transmission equipment and limits the performance of this equipment. This type of noise has a uniform distribution of energy over the frequency spectrum and a normal Gaussian distribution of levels. Equation 4–19 describes the thermal noise power relationship.

$$P_n = KTB_n \text{ watts (W)} \qquad (4\text{–}19)$$

where

$KT = 4.0 \times 10^{-21}$ W/Hz
$K = 1.380 \times 10^{-23}$ joules (J)/degree Kelvin (K)
$T = 17°C$ or 290 K
$B_n = $ noise bandwidth

A form used in communication more frequently is the noise relative to 1 milliwatt (mW), which is shown in equation 4–20.

$$P_n = 10 \log \frac{KTB_n}{1 \text{ mW}} \qquad (4\text{--}20)$$

Equations 4–21 and 4–22 depict voltage and current generators.

$$e_n^2 = KTBR \qquad (4\text{--}21)$$

$$I_n^2 = \frac{KTB}{R} \qquad (4\text{--}22)$$

Generally, voltage or current noise generators are more useful in electronic circuits, while power calculations are more useful for optical transmission. Noise calculations are of little use unless they are referenced to a signal. Therefore, an important relationship to consider is signal-to-noise ratio (SNR), which is a measure of channel or amplifier quality. SNR is depicted in Equation 4–23 in log form, which is the most useful.

$$\text{SNR} = 10 \log \frac{P_s}{P_n}, \quad \text{power ratio} \qquad (4\text{--}23)$$

The alternative forms in Equations 4–23 and 4–24 are useful in electronic calculations, but 4–23 is exclusively used in fiber optics.

$$\text{SNR} = 20 \log \frac{e_s}{e_n}, \quad \text{signal ratio} \qquad (4\text{--}24)$$

$$\text{SNR} = 20 \log \frac{I_s}{I_n}, \quad \text{signal ratio} \qquad (4\text{--}25)$$

When using SNR values in optical cabling systems, one must remember there is a difference. As an example, for a good video display the optical SNR is 40 dB and this would translate to an 80-dB SNR in an electrical system. Therefore, the designer must keep these differences in mind when dealing with the two technologies. As an added note, laser fiber-optic transmitters have difficulty in producing SNRs of 57 dB or greater, which implies that there is an upper bound in laser-driven systems.

Another important noise calculation that is particularly useful for evaluating amplifiers is the noise figure, given in Equation 4–26.

$$\text{NF} = \frac{\text{SNR}_{in}}{\text{SNR}_{out}} = \frac{S_i/N_i}{S_o/N_o}$$

$$F_{dB} = 10 \log \text{NF} = 10 \log \frac{N_o}{S_o N_i / S_i} \qquad (4\text{--}26)$$

$$= 10 \log \frac{N_o}{KTBG}$$

where

$$G = \text{gain} = \frac{S_o}{S_i}$$

In the equation, note that if noise out of the device is equal to noise in, the device is noiseless. An assumption made here is that no intermodulation products are present (i.e., the device is linear).

A noise attributed to the conversion of analog signals to digital form is called quantization noise. This type of noise is shown in Figure 4–11. It is due to quantization error, as shown in the figure. If an error is smaller than the quantization level, which is the usual case, generally it is less than one-half the amplitude. The quantization noise power is calculated next.

$$P(n) = \frac{1}{q} - \frac{q}{2} \leq n \leq \frac{q}{2}$$

$$= 0, \quad \text{otherwise}$$

$$\text{QNP} = \text{quantization noise power} = \int_{-q/2}^{q/2} \frac{n^2}{qR} \, dn$$

$$= \frac{1}{Rq}\left(\frac{2q^3}{3 \times 8}\right)$$

$$= \frac{q^2}{12R}$$

$$\text{SN}_q\text{R} = \text{signal to quantization noise ratio} \quad (4\text{–}27)$$

$$= 10 \log \frac{v^2}{q^2/12R}$$

$$= 10.8 + 10 \log \frac{v}{q}$$

where v is the analog signal level. A number of assumptions are made in Equation 4–27.

1. The quantization amplitude is constant, which is not the case for some types of adaptive A/D conversion. Some of the more advanced techniques reduce step size as the amplitude approaches overload. Overload is a condition when the signal amplitude is larger than the A/D converter can quantize. These adaptive A/D converters decrease step size when signals are low level. This adaptive encoding technique increases the dynamic range of the A/D converter.
2. The A/D converter has infinite rise and fall time (i.e., it is not band limited).

Idle channel noise is also present. If the noise at the input of an A/D converter is larger than the quantization level, the effect is a continuous string of pulses depicting the plus and minus values about zero. This type of noise will be

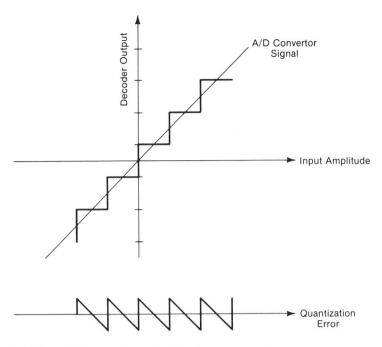

Figure 4-11 Quantization error for a single sloping analog input.

noticeable during pauses or some fricative sounds in speech. This idle noise may be reduced to zero if these low-level values are encoded as zeros. By selecting a quantization characteristic that will present a threshold to the analog input, the noise can be suppressed. The characteristic shown in Figure 4-11 is an example of this technique. Another possibility of course is to increase the quantization voltage levels, which will reduce sensitivity; this decision will depend on the application.

A different type of noise is due to fiber-optic parameters, occasionally referred as quantum noise. The first case to consider is a digital system where the number of detected photons per second is equal to the number of hole–electron pairs generated by incident light. Equation 4-28 will statistically predict the number of hole pairs generated; this is a Poisson distribution.

$$P_n(n) = \Lambda^n \frac{e^-}{n!} \qquad (4\text{-}28)$$

where

$$\Lambda = \frac{1}{hf} P_d(t)/dt = \frac{E_d}{hf}$$

E_d = energy detected over the time interval

If an assumption is made that no errors occur due to electron–hole pairs of energy E_d, then Equation 4–28 will have the following value:

$$P_n(0) = e^{-E_d/hf}$$

If, further, Equation 4–25 is evaluated for a BER of 10^{-10}, then the result is

$$\text{BER} = 10^{-10} = e^{-E_d/hf}$$
$$E_d = 23.3\ hf$$

The average number of photons in the optical pulse is 23.3. The average minimum power required to maintain this BER is as indicated in Equation 4–29.

$$P_{\min} = 23.3 \times \frac{1}{2} hf\ B_n$$
$$= 11.65\ hf\ B_n \qquad (4\text{–}29)$$

where B_n = noise bandwidth. The ½ is used because the digital waveform has a 50/50 duty cycle. BER is an indication of the SNR, but in digital systems it is more a figure of merit than SNR. When receiver calculations are made, dark current and thermal noise will also be considered.

To complete the optical noise calculations, the analog signal-to-noise ratio must be considered. The quantum-limited SNR, with thermal and dark current ignored, is calculated using Equation 4–30.

$$(\text{SNR})_q = \frac{m^2 P_o}{2hf\ B_n}$$
$$P_o = I_s^2 R_L G, \qquad I_s = P_o \gamma \qquad (4\text{–}30)$$
$$(\text{SNR})_q = \frac{m^2 (P_{\text{in}} \gamma)^2 R_L}{2hf\ B_n}$$

where
- γ = responsivity in amperes/watt (A/W)
- P_{in} = optical power at the detector
- R_L = detector load resistance
- m = modulation index

This equation represents the absolute limit of optical detection. Its derivation can be found in reference [4]. If thermal noise and dark current noise can be kept very small, this limit can be approached very closely.

In laser transmission of analog signals, the modulation index must be kept less than 60 percent (0.6) or the signals will be destroyed. From personal experience, 50 percent modulation will generally allow some margin of overdrive as a safety factor. Generally, when overdrive persists for less than 2 ms, the laser

Network Topology

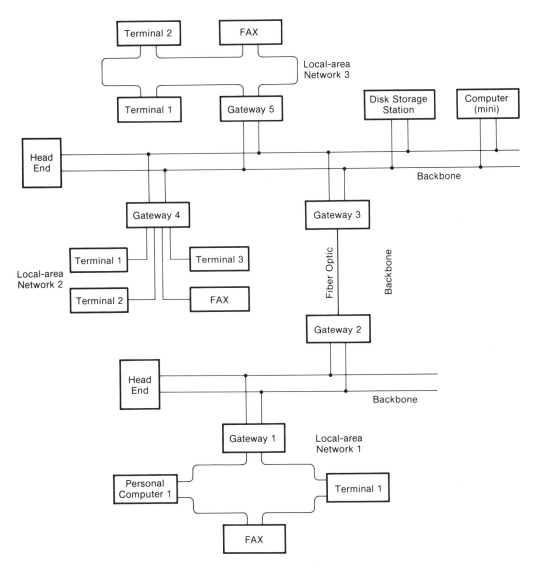

Figure 4-12 Three LANs connected by a backbone.

source will deteriorate to an unusable power output level or a catastrophic failure will occur.

Other sources of noise that are dependent on various applications will be discussed in Chapter 5.

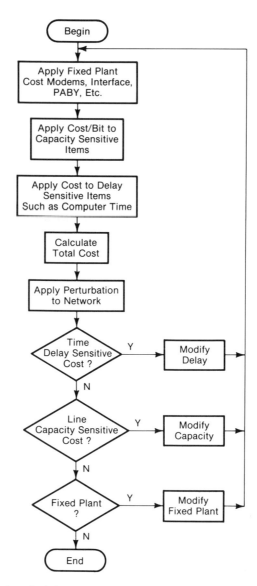

Figure 4–13 Flow chart depicting a method of finding cost of variables.

Backbone Design

The rudimentry design issues have been discussed previously. A major difference between backbone networks and other LANs is that they may connect several different LANs into one large network. The backbone network may have evolved

Network Topology

or it may be a planned architecture. A typical backbone network is shown in Figure 4-12. Note that this is a highly simplified backbone architecture, and the connection between gateways 2 and 3 could have been a telephone line. In this case, a severe limitation would have occurred, due to the low transmission rate possible for telephone facilities. Figure 4-12 is also an example of a system that may have evolved, rather than being of the planned variety. These backbone networks are usually quite massive and cost sensitive. The design methodology is usually trial and error.

One iterative method is as follows: a preliminary design is configured and is then tested for connectivity, delay, and other constraints, after which a cost analysis is completed. This network is then used for a baseline. The baseline network can be examined for improvements with a connectivity, delay, and cost analysis performed on each iteration. For large networks, a computer will be required to assist in these analyses. A flow chart of the iterative process is provided in Figure 4-13, which determines cost sensitivity of the network.

It is necessary to find the parameters responsible for causing the largest cost perturbations in the network. If these parameters are known, the designer may optimize the network more easily, or at least gain more insight when modifying the network. The modification blocks shown in the flow chart can be completed by the programmer or automatically inserted by the program. A plot can be designed into the program to examine total cost/Δ capacity (ΔC), total cost/Δ delay (Δd), and total cost/Δ fixed plant (ΔFC). Some of the computer-aided design (CAD) techniques for network analysis can be examined for possible use in this analysis. One problem with using CAD techniques is they are usually designed for linear analysis. Cost/ΔC and cost/ΔFC are usually nonlinear, but they can be considered linear for small excursions.

A number of rather sophisticated backbone design techniques are discussed in reference [5]. This subject is lightly treated here because the majority of design applications will be with LAN. Large backbone networks are usually designed with some of the previous facilities installed. This type of network will have preconditions due to spatial installations. A challenge to the designer's ingenuity will be to install a backbone under these conditions.

In Chapter 5 a network resembling a backbone network in some respects will be covered. It is the AT&T LAN with a centralized bus. This particular network combines the advantages of ring, bus, and star architectures. Chapter 6 will delve into the ISO layers of various networks that are common in today's marketplace.

PROBLEMS

4-1. Calculate the propagation delay for the following materials: (a) glass, (b) Mylar, (c) Teflon, (d) water, (e) ice.

4-2. Which of the materials listed in Problem 4-1 has the lowest delay and is it suitable for optical waveguides?

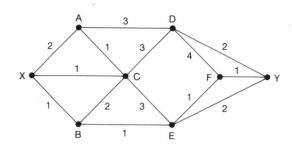

Figure P4-3 Network used in problem three of review questions.

4-3. Using network cuts, find the maximum possible flow across the network from X to Y in Figure P4-3.

4-4. Find the shortest path from X to Y. Refer to the graph in Figure 4-3(c).

4-5. If a total message must be received at a node before it can be repeated, derive the equation describing the transmission rate for a T1 line. A T1 transmission consists of a synchronization bit, 8 bits of quantization code, and 24 channels of audio sampled at the Nyquist rate. Use a 4-kHz audio bandwidth.

4-6. A transmission system uses one-half its capacity, total delay is 3 μs, and the message length is 256 bytes. What is the maximum transmission capacity in bits per second?

4-7. Find the maximum capacity of a video channel that has eight levels of coding, assuming standard 6-MHz bandwidth.

4-8. Given an optical receiver with a modulation index of 50 percent, responsivity of 0.5 A/W, R across the detector diode of 100, and a noise bandwidth of 300 MHz, find the quantum signal-to-noise ratio (-32 dBm optical input power). How does the noise power in the optical system compare with a copper cable?

REFERENCES

[1] A. S. Tanenbaum, *Computer Networks,* Prentice-Hall, Inc., Englewood Cliffs, N.J., 1981, p. 41.

[2] Ibid., pp. 54–55.

[3] C. E. Shannon, "A Mathematical Theory of Communication," *Bell System Tech. J.,* 27, pp. 379–423 (July 1948); pp. 623–656 (Oct. 1948).

[4] S. Personick, *Optical Fiber Transmission Systems,* Plenum Press, New York, 1981, pp. 90–93.

[5] Tanenbaum, op. cit., pp. 67–87.

5
PHYSICAL LAYER

The physical layer of the ISO model is the foundation layer of the network and it provides the absolute physical limitations for the LAN. Other limitations, which may be circumvented with program upgrades, can occur in the software. But cable plants have limitations such as bandwidth and loss per unit length that cannot be modified without undue expense.

The first part of the chapter is devoted to examining various performance criteria, such as waveform analysis and rise time. This section will acquaint the reader with various analysis techniques that are generally applicable to all cable plant designs.

Transmission error detection and corrections are examined. Some of the more popular error-detection integrated circuits are discussed. But most of the discussion will be of a general nature. Error correction is becoming more popular; computer manufacturers offer fault-tolerant memory boards, and in other segments of computer hardware, error-detection circuitry is present.

Telephone system coverage is included because many LANs use telephone facilities to remote LANs nodes and network segments. Multiplexing of signals is a necessity for telephone facilities, and this topic is addressed in this chapter. Both electronic and fiber-optic multiplexers and demultiplexers will be examined.

The next item to be considered is cable plant design. A study of both fiber-optic and copper cable plants is initiated with a performance comparison, but a price comparison cannot be made due to the plummeting costs of fiber-optic components. Some of the attributes of single-mode fiber optics are discussed because of their importance in long-haul networks. This discussion will give the reader a feel for what the future might hold in fiber optics.

Another subject to be discussed in the hierarchy is transmitters and receivers, of which several types are addressed. The discussion encompasses transmitter–receiver pairs for use with various cable plants, such as shielded twisted pairs, multimode fiber optic (digital baseband), multimode fiber-optic analog (carrier-type systems), baseband coaxial, and broadband coaxial. Transmitter and receiver codes for self-clocking waveforms are addressed in this section. A number of schematics are given showing various techniques of implementing a great deal of the aforementioned hardware.

Two final topics in this chapter are devoted to common applications of ring, star, and bus cable plants and the controllers that are available for implementing these technologies. A great deal of information is presented that should assist the would-be designer in making a choice of hardware to implement LAN designs. One must remember that this network will set the absolute limit, and design errors made here can be very costly.

Introduction to LAN Information Theory

Before any designs can be analyzed, the nature of the transmitted signals must be examined to determine what effects the medium has on waveforms. The waveforms shown in Figure 5–1 will be examined to determine their spectral plots. These waveforms are commonly found in baseband systems. The first waveform is described in the figure as NRZ (nonreturn to zero). This waveform has $+V$ as a binary one and $-V$ as a binary zero (i.e., the signal voltage never resides at zero) and will have an average signal of zero, if sufficient transitions are made over a sample interval. Equation 5–1 describes the summation of the spectral components required to construct a waveform, and for many waveforms the series quickly converges.

$$v(t) = \frac{a_0}{2} + \sum_{n=1}^{\infty} \left(a_n \operatorname{Cos} \frac{2n\pi t}{T} + b_n \operatorname{Sin} \frac{2n\pi t}{T} \right) \qquad (5\text{--}1)$$

The function shown in Figure 5–1a is considered an odd function; that is, $f(t) = -f(-t)$. Odd functions do not have a_0 and a_n components; the a component represents the dc term, and therefore the waveform has no dc component. This is important because it causes dc wander, which is discussed in the section on receivers. Equations 5–2, 5–3, and 5–4 are necessary to calculate coefficient values a_0, a_n and b_n.

$$a_0 = \frac{2}{T} \int_0^T f(t)\, dt \qquad (5\text{--}2)$$

$$a_n = \frac{2}{T} \int_0^T f(t) \operatorname{Cos} \frac{2n\pi t}{T}\, dt \qquad (5\text{--}3)$$

$$b_n = \frac{2}{T} \int_0^T f(t) \operatorname{Sin} \frac{2n\pi t}{T}\, dt \qquad (5\text{--}4)$$

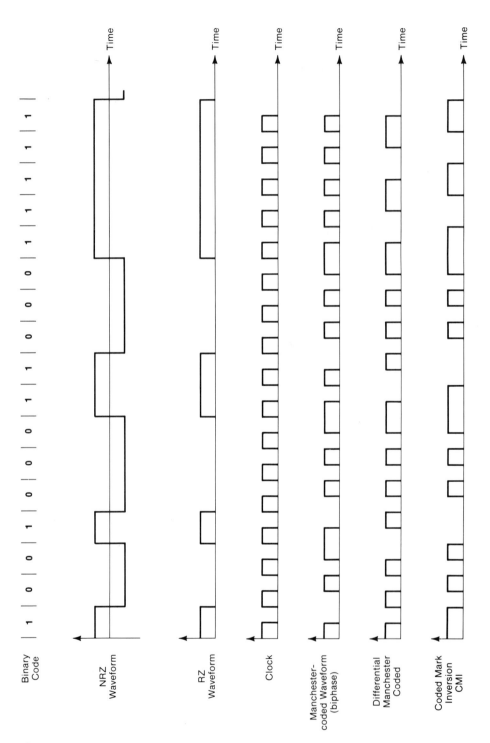

Figure 5–1 A common baseband waveform.

The calculations for the sample NRZ waveform shown in Figure 5–2(a) are:

$$a_0 = a_n = 0 \text{ for odd functions}$$

$$b_n = \frac{2}{T}\int_{-T/2}^{0} -A \sin\frac{2n\pi t}{T} dt + \frac{2}{T}\int_{0}^{T/2} A \sin\frac{2n\pi t}{T} dt$$

$$= \frac{2}{T}\frac{TA}{2\pi n}\cos\frac{2n\pi t}{T}\bigg]_{-T/2}^{0} - \frac{2}{T}\frac{TA}{2\pi n}\cos\frac{2n\pi t}{T}\bigg]_{0}^{-T/2}$$

$$= \frac{2A}{\pi n}(1 - \cos n\pi)$$

$$= \frac{4A}{\pi n}\sin^2\frac{n\pi}{2}$$

$$v(t) = 2A\sum_{n=1}^{\infty}\frac{\sin^2(n\pi/2)}{(n\pi/2)}\sin\frac{2n\pi T}{T} \quad (5\text{-}5)$$

$$= 2A\left[\frac{2}{\pi}\sin\frac{2\pi t}{T} + \frac{2}{3\pi}\sin\frac{6\pi t}{T} + \frac{2}{5\pi}\sin\frac{10\pi t}{T} + \ldots\right]$$

$$= \frac{4A}{\pi}\sum_{n=0}^{\infty}\frac{1}{2n+1}\sin\frac{2(2n+1)\pi t}{T}$$

Equation 5–5 can be plotted as in Figure 5–2(a); this is a spectral amplitude plot of the NRZ sample. Whenever the waveform is such that a_0 terms appear, a dc value is present. If waveforms are not odd, then they may have a dc component. Waveforms that are NRZ, which have large strings of ones or zeros, will have a dc component and all RZ (return to zero) waveforms have a dc component. But if RZ waveforms are ac coupled to the receiver, then dc wander can be minimized.

The next example will be the situation when large strings of zeros occur and a single one appears in an NRZ waveform [see Figure 5–2(b)]. The function shown is an even function; that is, $f(t) = f(-t)$. The calculations for the spectral analysis are as follows:

$$b_n = 0, \quad \text{for all } n \text{ because the function is even, i.e., } f(t) = f(-t)$$

$$a_0 = \frac{2}{T}\int_{-T/2}^{T/2} f(t)\, dt$$

$$= \int_{-T/2}^{T/2} A\, dt \quad f(t) = 0, \quad t > \frac{\tau}{2}, t < -\frac{\tau}{2}$$

$$= \frac{2\tau}{T}A$$

$$a_n = \frac{2}{T}\int_{-T/2}^{T/2} A \cos\frac{2n\pi t}{T}\, dt$$

$$= \frac{2}{T}\left(\frac{T}{2n\pi}\right)\left(-\sin\frac{2n\pi t}{T}\right)\bigg]_{-\tau/2}^{\tau/2}$$

$$= -\frac{2}{n\pi}\sin\frac{2n\pi\tau}{T}$$

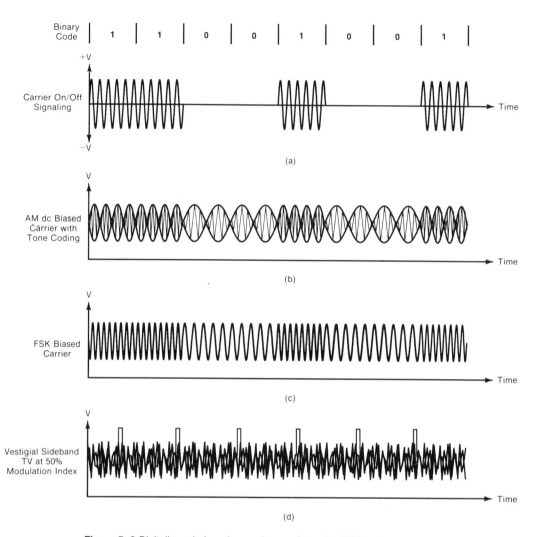

Figure 5–2 Digitally coded analog carriers and standard TV analog signals.

Substitute a_0 and a_n in Equation 5–1; the result is shown in equation 5–6.

$$v(t) = \frac{\tau A}{T} - \sum_{n=1}^{\infty} \frac{2}{n\pi} \operatorname{Sin} \frac{2n\pi\tau}{T} \operatorname{Cos} \frac{2n\pi t}{T} \qquad (5\text{–}6)$$

Equation 5–6 is a general equation which can be used for evaluation of waveforms that begin to approach an impulse function. The absolute value for the a_n coefficient can be put into the Sin x/x format as follows:

$$|a_n| = \frac{2A}{n\pi} \text{Sin} \frac{2n\pi\tau}{T} = \frac{4\tau A}{T} \frac{\text{Sin } n\omega\tau}{n\omega\tau}$$

As $T \to 0$ the Sin x/x ratio will become 1 for all n, and the function $v(t)$ will have a flat spectrum at all frequencies.

The desirable type of transmission waveforms should have transitions to prevent a single bit from resembling an impulse. As previously mentioned, dc wander is also a problem. Let us now examine the equations using the values in Figure 5–2(b) for T.

$$T = 5 \text{ bit times}$$
$$\tau = 1/5T \qquad (5\text{–}7)$$
$$v(t) = \frac{A}{5} - \sum_{n=1}^{\infty} \frac{2A}{n\pi} \text{Sin} \frac{2n\pi}{5} \text{Cos } n\omega t$$

The dc term and the negative term were neglected in figure 5–2(b) for ease of plotting. Plots of the Sin x/x form are frequently found in mathematics and electronics information theory books.

A technique of forcing transitions in the waveforms to reduce dc wander and prevent impulse functions from occurring is known as bit stuffing. This technique is used by Western Digital in their WD1933 integrated circuit controller. The waveform is shown in Problem 5–1 (Figure P5–1). Note the zero inserted after five consecutive ones.

Thus far the only type of transmission considered has been digital baseband. Analog signals must now be considered. Figure 5–3 is a diagram of commonly used analog transmission waveforms.

The first technique is on/off keying (OOK) of a carrier. If a binary pulse train is superimposed on a carrier using OOK, then the baseband spectral plot will be shifted up in frequency, as shown in Figure 5–4(a) and (b). The proof of this statement is left for the reader in Problem 5–3.

The next waveform to be considered is an AM tone-modulated carrier with a large dc offset, as shown in Figure 5–3(b). The reason for the large offset is due to laser modulation. Lasers cannot be modulated at 100 percent intensity excursions, as will be shown in the section on transmitters, or the waveforms will become distorted and damage to the laser facet will result. The waveform shown in the figure is similar to the laser optical signal. The offset is due to laser biasing. A spectral plot of this signal is shown in Figure 5–5. Note that it resembles a standard AM plot except a large dc component is present due to the dc offset. The bandwidth required to reproduce this signal is $2\pi B$ instead of B, the original digital bandwidth.

The next type of modulation is shown in Figure 5–3(c). This is perhaps one

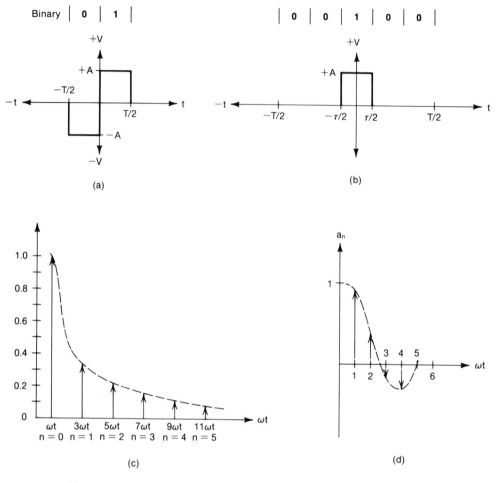

Figure 5–3 (a) NRZ odd function sample: $f(t) = -f(-t)$. (b) RZ even function sample: $f(t) = f(-t)$. (c) Spectral plot of Figure 5–2(a). (d) Spectral plot of terms inside the summation of Equation 5–7 and ignoring the dc term.

of the most important types. It is widely used as a modulation technique on broadband networks and in telephone modem technology. As one can observe, it is a frequency modulation (FM) technique. The technique used is frequency shift keying (FSK).

Two types of FM will be investigated in this section: narrow band and wide band. They are defined as follows:

$$\Delta f \ll B$$
$$\Delta f \gg B$$

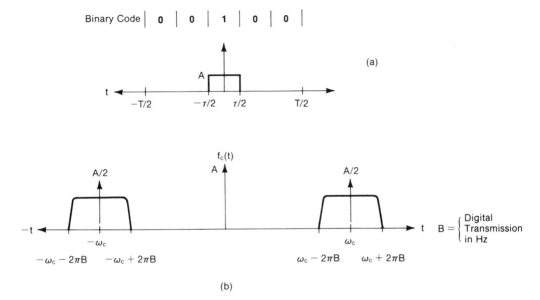

Figure 5–4 (a) Baseband binary coded signal. (b) On/off keying with spectral plot shifted up in frequency.

where Δf = frequency spacing or deviation. The first expression is used when the spacing between the two FSK frequencies is small compared to the digital bandwidth of the binary modulating signal. The second expression is used when the spacing between the two shifted frequencies is large compared to the bandwidth. Telephone modems using switched networks are an example of the first case, and the second case is apparent in some broadband FSK modems. For the first case, the transmission bandwidth will approach $2B$, and for the latter case, $2\Delta f$.

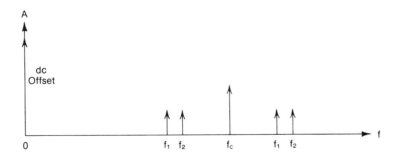

Figure 5–5 Spectral plot of a tone-modulated AM signal with a dc offset.

The deviation is a function of the voltage difference between a binary 1 and 0 in digital FSK systems and the sensitivity of the FM modulator.

$$f_c(t) = \text{Cos}\left(\omega_c t + \frac{\Delta f}{B} \text{Sin } \omega_B t\right) \qquad (5\text{--}8)$$

Equation 5–8 describes an FM sinusoidally modulated carrier.
Using the trigonometric identity Cos $(A + B)$,

$$f_c(t) = \text{Cos } \omega_c t \text{ Cos}\left(\frac{\Delta f}{B} \text{Sin } \omega_B t\right)$$
$$- \text{Sin } \omega_c t \text{ Sin}\left(\frac{\Delta f}{B} \text{Sin } \omega_B t\right) \qquad (5\text{--}9)$$

$$\frac{\Delta f}{B} << \frac{\pi}{2}, \qquad \text{narrowband FM restriction}$$

Then

$$\text{Cos}\left(\frac{\Delta f}{B} \text{Sin } \omega_B t\right) \approx 1$$

$$\text{Sin}\left(\frac{\Delta f}{B} \text{Sin } \omega_B t\right) \approx \frac{\Delta f}{B} \text{Sin } \omega_B t, \qquad \text{for the sine of small angles}$$

$$f_c(t) \cong \text{Cos } \omega_c t - \frac{\Delta f}{B} \text{Sin } \omega_c t \text{ Sin } \omega_B t \qquad (5\text{--}10)$$

$$= \text{Cos } \omega_c t - \frac{\Delta f}{2B} [\text{Cos } (\omega_c - \omega_B)t + \text{Cos } (\omega_c + \omega_B)t]$$

For the narrowband case, note that the equation is similar to AM modulation. The bandwidth is twice the digital bandwidth in hertz, or $2B$.

Next, the general case will be considered, that is, wideband FM with modulation index $\Delta f/B > \pi/2$. The Sine and Cosine series are used to expand the two parenthetical expressions in Equation 5–11. Let $\Delta f/B = \beta$, the FM modulation index. Then

$$f_c(t) = \text{Cos } \omega_c t \text{ Cos } (\beta \text{ Sin } \omega_B t) - \text{Sin } \omega_c t \text{ Sin } (\beta \text{ Sin } \omega_B t) \qquad (5\text{--}11)$$

Cosine series expansion:

$$\text{Cos } (\beta \text{ Sin } \omega_B t) = 1 - \frac{\beta^2 \text{ Sin}^2 \omega_B t}{2!} + \frac{\beta^2 \text{ Sin}^4 \omega_B t}{4!} + \ldots$$

Sine series expansion:

$$\text{Sin}(\beta \, \text{Sin} \, \omega_B t) = \beta \, \text{Sin} \, \omega_B t - \frac{\beta^3 \, \text{Sin} \, \omega_B t}{3!} + \ldots$$

$$f_c(t) = \left(1 - \frac{\beta^2}{4} + \frac{\beta^4}{96} + \ldots\right) \text{Cos} \, \omega_c t + \frac{\beta}{2}\left(1 + \frac{\beta^2}{6} - \ldots\right) \quad (5\text{–}12)$$

$$[\text{Cos}(\omega_c + \omega_B) - \text{Cos}(\omega_c - \omega_B)t] + \frac{\beta^2}{8}\left(1 + \frac{\beta}{3}\right)$$

$$[\text{Cos}(\omega_c + 2\omega_B)t + \text{Cos}(\omega_c - \omega_B)t] + \ldots$$

Equation 5–12 is the result of the series substitutions into Equation 5–11. Obviously, then, the expression for large modulation index is rather difficult to evaluate. Fortunately the values can be determined quickly using Bessel functions of the first kind. Many texts have either curves or tables of these functions. Equation 5–12 written in Bessel function form is given by Equation 5–13.

$$f_c(t) = A_c\{J_0(\beta)\text{Cos} \, \omega_c t - J_1(\beta)[\text{Cos}(\omega_c - \omega_B)t - \text{Cos}(\omega_c + \omega_B)t]$$
$$+ J_2(\beta)[\text{Cos}(\omega_c - 2\omega_B)t + \text{Cos}(\omega_c + 2\omega_B)t] \quad (5\text{–}13)$$
$$+ J_3(\beta)[\text{Cos}(\omega_c - 3\omega_B)t + \text{Cos}(\omega_c + 3\omega_B)t]$$
$$+ J_4(\beta)[\text{Cos}(\omega_c - 4\omega_B)t + \text{Cos}(\omega_c + 4\omega_B)t] + \ldots\}$$

Table 5–1 can be used to find the amplitude of various harmonics for a given modulation index B. For modulation indexes of less than 1, use Equation 5–12 with the approximation equation (5–14).

**TABLE 5–1
Short Table of Bessel Functions**

B/Jm	J_0	J_1	J_2	J_3	J_4	J_5	J_6	J_7	J_8	J_9	J_{10}	J_{11}	J_{12}
1	0.8	0.45	0.10	—	—	—	—	—	—	—	—	—	—
2	0.2	0.60	0.35	0.15	0.03	—	—	—	—	—	—	—	—
3	−0.30	0.30	0.50	0.30	0.12	0.05	—	—	—	—	—	—	—
4	−0.40	−0.005	0.38	0.42	0.30	0.10	.05	—	—	—	—	—	—
5	−0.18	−0.3	0.005	0.38	0.40	0.30	0.10	0.05	—	—	—	—	—
6	0.18	−0.28	−0.22	0.12	0.35	0.35	0.25	0.10	0.05	—	—	—	—
7	0.30	−0.02	−0.30	−0.18	0.18	0.35	0.35	0.23	0.12	0.05	0.01	—	—
8	0.15	0.22	−0.10	−0.30	−0.10	0.18	0.35	0.22	0.10	0.05	0.05	0.02	—
9	−0.10	0.22	0.17	−0.18	−0.28	−0.05	0.20	0.35	0.30	0.20	0.12	0.05	0.03
10	−0.22	0.02	0.22	0.05	−0.20	−0.22	0	0.20	0.30	0.30	0.20	0.12	0.08
11	−0.18	−0.20	0.10	0.20	−0.03	−0.22	−0.20	0	0.20	0.30	0.28	0.20	0.12
12	0.14	−0.22	−0.10	0.20	−0.18	−0.10	−0.25	−0.12	0.05	0.22	0.30	0.28	0.20
13	0.20	−0.10	−0.20	0	0.20	0.12	−0.10	−0.22	−0.15	0.05	0.22	0.30	0.25
14	0.18	0.10	−0.16	−0.18	0.10	0.20	−0.12	−0.18	−0.22	−0.12	0.10	0.22	0.30
15	−0.03	0.20	0.03	−0.18	−0.10	0.10	−0.20	0.03	−0.20	−0.20	−0.10	−0.10	0.28

$$f_c(t) \approx \left(1 - \frac{\beta^2}{4}\right)\cos\omega_c t + \frac{\beta}{2}[\cos(\omega_c + \omega_B)t - \cos(\omega_c - \omega_B)t]$$
$$+ \frac{\beta^2}{8}[\cos(\omega_c + 2\omega_B)t + \cos(\omega_c - 2\omega_B)t] \quad (5\text{-}14)$$

The B^2 terms will be smaller than 12.5 percent of the carrier amplitude; they may be dropped depending on the precision required in the calculations, which reduces the calculations to narrowband FM: $\Delta f/B < \pi/2$. A spectral plot is provided in Figure 5–6(a) and (b) showing narrowband and wideband FM examples. The plots are absolute values taken from Table 5–1.

The bandwidth in Figure 5–6(a) is approximately $2B$ because the amplitude of $F_c \pm 2B$ sidebands is approximately 10 percent of the unmodulated carrier amplitude, which may be neglected. The justification for this rule of thumb can be found in reference [1]. For narrowband FM, the bandwidth (BW) is approximately $2B$. Wideband FM [Figure 5–6(b)] is $12B$ or 12 times the highest transmission rate in hertz.

$$\beta = \frac{\Delta f}{B}, \quad 2\Delta f = \text{BW} = \text{FM modulation index}$$

Then $\Delta f = 6B$.

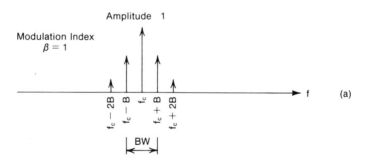

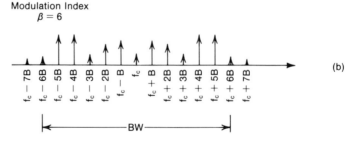

Figure 5–6 (a) FM with small modulation index, $b = 1$. (b) FM with large modulation index.

A typical modem used for transmission of data over telephone-switched networks is the 202 type. The tones are as follows (the modulation index indicates the modem operates in the narrowband FM mode):

    Mark   2200
    Space   1200
    Transmission   1200 baud
    Bandwidth   2500 Hz

$$\beta = \frac{\Delta f}{B} = \frac{500}{1200}, \quad 2\Delta f = 1000 \text{ Hz}$$
$$= 0.41 < \frac{\pi}{2}$$

Sytek produces a frequency agile modem that selects one of 20 channels within a 6-MHz group. Channel spacing is 6 MHz per 20 or 300 kHz. The frequency deviation Δf is $= 35$ kHz, with a maximum transmission rate equivalent to 19.2 kHz.

$$\beta = \frac{\Delta f}{B} = 1.8, \quad B = 19.2 \text{ kHz}$$

Examine Table 5–1. At least three sidebands are needed to pass the FM with good fidelity. Therefore, the bandwidth should be $6B$ or 115.2 kHz. For a more comprehensive study of FM theory, reference [1] gives a wealth of information on the subject.

A topic of concern to the baseband LAN designer is the rise and fall time of pulses. As pulses become "smeared" or stretched, they become more difficult to detect. This condition leads to a high bit error rate (BER) value, which is a performance parameter of the network.

Both copper and fiber-optic cable rise times will be presented, with the greatest emphasis on fiber optics. Many technologists believe that by the mid to late 1990s very little copper cable will be implemented in new cable plant designs. However, anyone acquainted with the transition from electronic tubes to transistor technology will recall that for many years tubes lingered on after the transition. All the copper cable will obviously not be replaced immediately (e.g., copper plants installed today are expected to remain intact for 20 to 30 years depending on the installation).

For copper cable, it is the capacitance per unit length that affects the pulse rise time. Rise time is also related to bandwidth, as shown in Equations 5–15 and 5–16.

$$\text{Bandwidth (BW)} = \frac{1}{\tau} \quad (5\text{–}15)$$

Equation 5–15 will allow the signal to pass, but the distortion will make it not look very rectangular. It will more closely resemble a triangular pulse. If pulse fidelity were not important, the BW calculation in the equation would be adequate. But for baseband transmission, the received pulse should be a good replica of the transmitted pulse. Equation 5–16 describes the bandwidth pulse width relationship that is the most desirable, and the rise time from 0 to 100 percent amplitude is given by Equation 5–17. This makes the assumption that the waveform passes through the system with no bandwidth restrictions (i.e., BW $\gg$ 1/τ).

$$BW = \frac{5}{\tau} \qquad (5\text{-}16)$$

$$\tau_{RT} \approx \frac{0.8}{BW} \qquad (5\text{-}17)$$

In copper systems, the bandwidth is affected by the cable capacitance, signal, and ground paths in the cable, whereas in fiber-optic cable plants the rise time is affected by dispersion. To examine both phenomena in detail, specific cable plants should be investigated.

Three types of copper cable plants will be considered although many more exist. The first is the twisted shielded pair, which is generally used for low-frequency baseband transmission. The transmission rates are generally less than 2 Mbits/s. The second type of cable plant is similar to Ethernet (i.e., high-transmission-rate baseband type with transmission rates from 2 to 10 Mbits/s). The third is the broadband system using coaxial cable, such as Wangnet or Sytek. The coverage of copper cable plants will not be as detailed as for fiber-optic plants because the technology has been used for many years and is well documented. Fiber optics, on the other hand, has many new components and techniques, which are, in many cases, still in the laboratory.

Cable Plant Design

When installing taps on copper cable plants, tuning may be required to prevent mismatches that cause line reflection and large standing waves on the cable. Most LAN vendors provide instructions for adjusting taps. However, for the design of matching network, the reader can consult references [2] and [3]. Many techniques use a tapping method that can be accomplished without tuning.

One tapping method requires the use of power splitters. These units are matched to the line. The main transmission line must be cut and connectors installed at the break in the line; then the power splitter is inserted. Note that there are three problems with this type of tapping technique: (1) service must be disrupted when a tap is being installed, (2) the tap will have a large insertion loss of at least 3dB, and (3) large numbers of repeaters are necessary. Another type of tap has high impedance, which allows the cable plant to be tapped without disruption in service, such as in Ethernet.

The first discussion of a copper cable plant will deal with a two-wire twisted and shielded pair. The cable plant will consist of a ring that can be either token passing (CSMA [Carrier Sense Multiple Access] or CSMA/CD [Collision Detect]). A diagram of the ring is provided in Figure 5–7(a). The nodes are provided with fail-safe relays that bypass the node should a power outage occur (i.e., the terminal connected to the node may cause the power cord to be released from the outlet or a fuse may be blown). The topology is used for connecting teletypes together in a 20 or 80 milliampere (mA) loop; this technique is also used for interconnecting word processors. The node interface is shown in Figure 5–7(b) with the fail-safe relay wiring. The relays are powered as shown in the figure. Power can be controlled to the relay with logic circuits to allow relay dropout if other machine malfunctions occur, such as a faulty transmitter or receiver. Relay dropouts can be controlled by the operator with break commands or by switching the terminal to local mode, which will allow message preparation off line.

A common 20-mA loop transmitter and receiver are shown in the schematic of Figure 5–7(c). Circuit functions are described in the following paragraphs.

The transmitter will be considered first. Two signals are shown: input, which is the binary or encoded data stream, and TEN (transmitter enable). This signal will enable or disable the transmitter with a logic one or zero, respectively. The gate U_1 drives Q_1, which controls the two current sources Q_2 and Q_3. When Q_1 is on 22 mA, current flow is established through R_3, R_4, and R_5. This current flow establishes a $+4$-volt (V) level at the base of Q_2 and a -11-V level at the base of Q_3. With Q_2 and Q_3 on, 22 mA of current is drawn from the $+5$-V transmitter supply, and this current enters the -15-V supply. Note that current always flows in both wires of the circuit which are equal and opposite in direction. The signals are pseudodifferential (i.e., they are independent but they are equal), and crosstalk is lower in this circuit than in a single-ended transmitter. Another important feature, capacitance, is present. The current sources will limit the charging rate and thus prevent large switching transients on the line.

The companion receiver is also shown in Figure 5–7(c). The network shown ahead of the receiver is for matching purposes and filtering. Capacitance is required in the network when it is connected to older model 33 or 27 teleprinters. This equipment has mechanical contacts for switching, which cause noise. Q_4 and Q_5 have biasing resistors as shown. The bases of Q_4 and Q_5 have $+2$- and $+13$-V bias, respectively. When the transmitter current sources are switched on, 22 mA will flow through the collector circuits of Q_4 and Q_5. This current flow will produce a voltage drop across the emitter resistors of Q_4 and Q_5 (1.9 V), thus turning on Q_6 and Q_7. Both transistors Q_6 and Q_7 must switch to change the state of the latch circuits. The switching speed of this circuit is generally limited by cable capacitance. A discussion of the performance will be presented after examining cable plant design aspects.

Figure 5–7(d) depicts an optical-coupler approach to the 20-mA loop. Note that the 22-mA current is only supplied by the transmitter. The receiver is completely isolated from the transmitter by the optical coupler. The optical coupler must be provided with some form of current limit, such as the zener and resistor shown connected to the diode of the optocoupler. If the wiring is connected in reverse, the zener will protect the optocouple from damage.

When the transmitter transistors Q_2 and Q_3 are turned on, as described in the previous circuit of Figure 5–7(c), 22 mA will flow through the coupler diode D_c, which will turn on the optocoupler transistor. Approximately 4 mA of current will flow through R_{10}, R_{11}, and R_{12}; this will turn on Q_6 and Q_7. The circuit will thus operate similarly to the receiver of Figure 5–7(c). In this circuit, the cable is driven differentially (i.e., the two current sources are dependent). The currents through both wires are equal and opposite; this increases crosstalk performance over the previous circuit.

An aspect of design that will not be considered is that of cable installed outside buildings. For installations outdoors, voltage-variable resistors or varistors must be installed at receivers and transmitters to protect electrical equipment from environmentally induced transients, which will damage equipment and present a shock hazard. Metal oxide varistors (MOVs) are available for this purpose. Optical couplers can be purchased without sufficient isolation to isolate receivers to comply with FCC part 68 regulations. This is a federal regulation dealing with telephone communication line protection. For operating in a safe manner, one may want to consider these regulations, but compliance is not mandatory unless connections are made to telephone facilities. Some optocouplers meet the regulations, but of course they are more expensive.

A large number of interface integrated-circuit receivers and transmitters comply with the Electrical Industries Association (EIA) standards. Complying with these standards allows the equipment to be flexible (i.e., to interface with other manufacturers' equipment). ICs such as the 9616 EIA RS-232-C/MIL-STD-188c, 9636A EIA 423 drivers, 9627 EIA RS-232-4 MIL-STD-188c, 9637 EIA RS-422 and RS-423 line receivers are examples of components that comply with these standards. Most integrated-circuit manufacturers list these circuits in the interface circuits section of their specifications book.

Thus far the discussion has dealt with baseband low-baud-rate circuits. The next item to consider is baseband cable plants that use coaxial cable, such as Ethernet. The transmission rates are from 2 to 20 Mbits/s. A slightly different class of problems exists with this technology. Some of the physical parameters will be investigated prior to discussing the functionality of the block diagrams.

Ethernet goals are to keep connections simple and low-cost; to keep circuit and protocol compatibility intact by avoiding options that promote incompatibility; the network should have addressing flexibility; all nodes should have equal network access; destruction or failure of one node should not cause catastrophic failure; the network should operate at high speed with low amounts of delay; the network should be configured with ease of maintainability; and the network should allow a layered type of architecture.

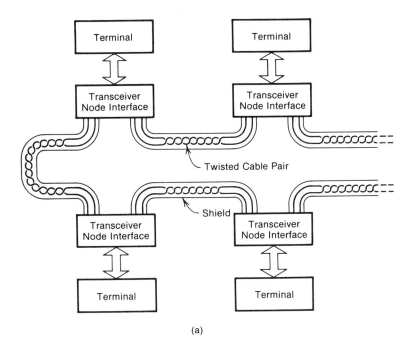
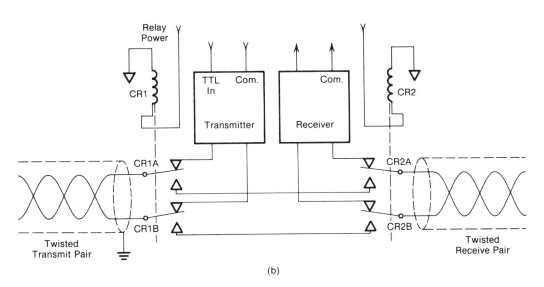

Figure 5–7 (a) Ring network implemented with shielded twisted pair cable. (b) Node interface designed with fail-safe relays. (c) Low data rate twisted shielded pair with transmitter and receiver details. (d) Low data rate twisted shielded pair designed with optocoupler isolation between transmitter and receiver.

(Continued on following pages)

Figure 5–7 (Continued)

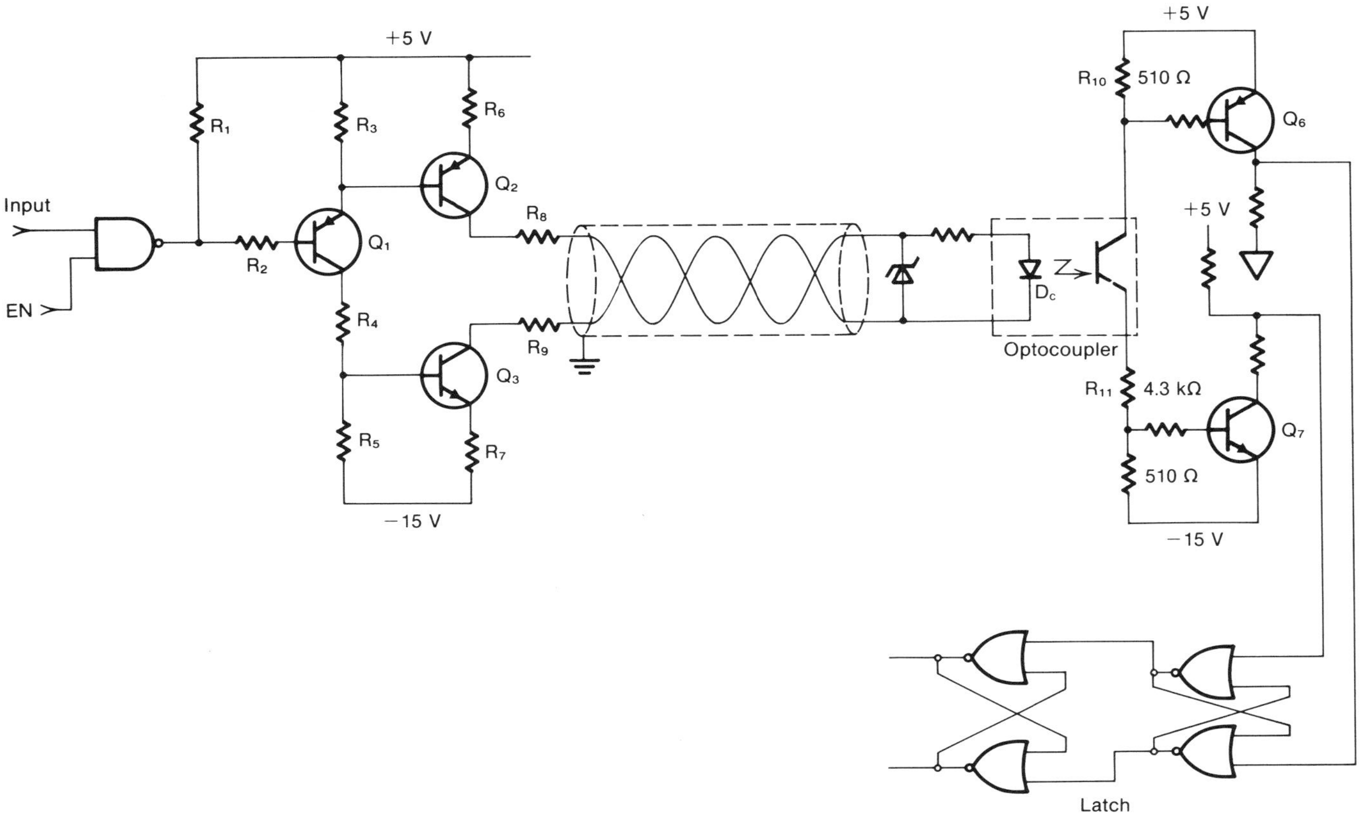

Figure 5-7 (Continued)

Ethernet Physical Layer Characteristics

Coaxial cable plants provide a medium for transmission of data. The cable plant is composed of segments that have a maximum length of 500 m; they are terminated in the characteristic impedance of the 50-Ω cable. A maximum of two repeaters between any two stations can be implemented. The repeaters may be located within a segment or at the end of a segment. Repeaters can be used to extend the topology (i.e., repeaters can drive segments to the allowed length of the cable plant). Repeaters occupy a transceiver position on the cable; therefore, they reduce the maximum station count to 1028.

A simple one-segment cable plant is shown in Figure 5–8(a). It has no repeaters. Figure 5–8(b) is a two-segment network with a repeater connecting them. Figure 5 8(c) shows a large-scale network; note that repeater 3 is connected in the center of the segment with stations 2 and 5 (i.e., it shows repeater flexibility).

Some of the important restrictions on an Ethernet cable plant are that the end-to-end maximum length is limited to 2.5 km or less and the longest path between transceivers is less than or equal to 1.5 km on the coaxial cable. The minimum distance between any two transceivers must be equal to or greater than 2.5 m.

Another parameter of great importance is coaxial cable delay. Propagation velocity in Ethernet cable is $0.77c$ ($c = 3 \times 10^5$ km/s, the speed of light in a vacuum). Round-trip delay is calculated using Equation 5–18.

$$\tau_E = \frac{2 d_m}{0.77c} \qquad (5\text{--}18)$$

where d_m = maximum distance between transceivers.

$\tau_E \approx 13$ μs   (worst-case)

A delay due to the transceiver twisted pair must also be considered. Worst-case propagation velocity is $0.65c$, with a maximum cable length of 50 m. Longer cables may be used, but they are special coaxial or fiber-optic cable. The worst-case round trip for twisted pair is calculated using Equation 5–19.

$$\tau_T = \frac{6 d_{mT}}{0.65 c} \qquad (5\text{--}19)$$
$$= 3.08 \text{ μs} \quad \text{(worst-case)}$$

The six is required in the expression because two repeaters can be implemented between the two communicating transceivers. Delays are very important when deciding packet length and for calculating efficiency, which is discussed in Chapter 7.

A propagation delay calculation due to physical parameters is shown in Table 5–2. The figures shown are all for worst-case conditions for round-trip delay.

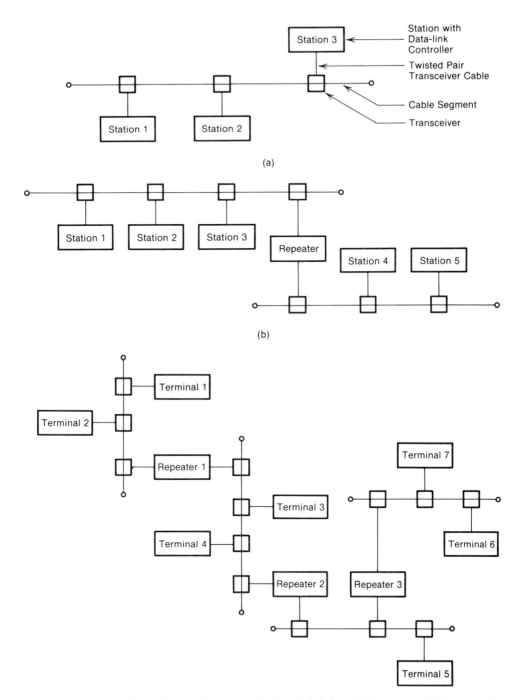

Figure 5–8 (a) Small-scale Ethernet cable plant installation. (b) Two-segment Ethernet cable plant with a repeater connecting them. (c) Large-scale Ethernet cable plant using multiple segments and repeaters.

**TABLE 5–2
Ethernet Physical Channel Propagation Delay**

Delay Element	Unit Steady-state Delay	Unit Start-up Delay	Forward Path	Return Path	Total Delay
Encoder	0.1 μs	0	3 × 0.1 μs	3 × 0.1 μs	0.60 μs
Transceiver cable	5.13 ns/m	0	300 m (1.539 μs)	300 m (1.539 μs)	3.078 μs
Transceiver transmit path	0.50 μs	0.2 μs	3 × 0.1 μs	3 × 0.1 μs	1.30 μs
Transceiver receive path	0.50 μs	0.5 μs	3 × 0.1 μs	0	1.30 μs
Transceiver collision path		0.5 μs	0	3	1.50 μs
Coaxial cable	4.13 ns/m	0	1500 m (6.5 μs)	1500 m (6.5 μs)	13.00 μs
Point-to-point cable	5.13 ns/m	0	1000 m (5.1 μs)	1000 m (5.1 μs)	10.26 μs
Repeater path	0.8 μs	0	1.6 μs	0	1.6 μs
Repeater collision	0.2 μs	0	0	0.4 μs	0.4 μs
Decoder	0.1 μs	0.8 μs	0.9 μs	0	1.8 μs
Carrier sense	0	0.2 μs	3 (0.6 μs)	0	0.6 μs
Collision detect	0	0.2 μs	0	3 (0.6 μs)	0.6 μs
Signal rise time 0–70%	0	0.1 μs	3 (0.3 μs)	0	0.3 μs
Signal rise time 50% to 94%	0	2.7 μs	0	3 (8.1 μs)	8.1 μs

Worst-case round-trip delay: 44.45 ns

This table will be referred to in other sections of the book where some of the entries will be discussed.

The next topic is the transceivers, which must be connected very close to the coaxial cable and must be less than or equal to 3 cm. A schematic of the transceiver is provided in Figure 5–9(a). The transceiver is composed of four elements, as shown in the diagram. The shunt resistance presented to the cable by the transceiver must be greater than or equal to 100 kilohms (kΩ), with a shunt capacity less than or equal to 2 picofarads (pF).

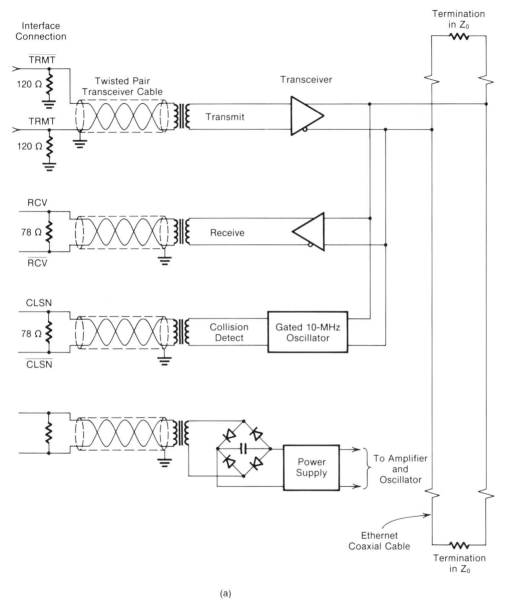

(a)

Figure 5-9 (a) Ethernet transceiver and cable diagram. (b) Integrated-circuit Ethernet interface and controller schematic.

(Continued on following page)

Figure 5–9 (Continued)

The transmitter and receiver are self-explanatory, but as shown in the diagram a collision-detection circuit is present. The logic within this unit must assert a collision within two bit times (200 ns). If transitions are sensed on the receiver twisted pair or the collision-detect pair, the signal is ORed and processed within the interface. Following the loss of carrier information, the channel must deassert the carrier signal within 160 ns.

Power is also provided to the transceiver by the interface unit. The station interface must supply a dc voltage between +12 and +15 Vdc at 0.5 A to a converter, which is not shown in the diagram.

The signaling current on the coaxial cable produced by the transceiver has the following specifications: ac components must be between ±14 mA and I_{dc}, where I_{dc} is the average dc component of the signal. This dc component must have a nominal value of 20 mA, and the dc component includes timing distortion effects. The maximum I_{dc} is 24 mA and a minimum value for this component is 18 mA. The signal's current is important. When two or more transmitters try to transmit simultaneously, the I_{dc} level increases; this indicates a collision has taken place. The collision detector only need sense this level change to trigger.

The specifications require that the transceiver draw bias current between −2 and 25 A in the off and on states. This is one of the important parameters. The specifications for Ethernet coaxial cable plants is much too extensive to describe in this book. Only the highlights of the specifications have been presented here. The specifications contain cable outside diameter, types of terminators, jacketing markings, connectors, types to be used, and the like. For further information, see reference [4].

The previous discussion considered the connections between the coaxial cable and the transceiver. The next connection to be considered is the transceiver to interface [see Figure 5–9(a)]. The connections consist of four shielded twisted pairs between the transceiver transmitter, receiver, collision detect, and power supply and the station interface circuitry. Figure 5–9(b) shows the details of an integrated-circuit (IC) version of an Ethernet serial interface (ESI). It is produced by the Intel Corporation as ESI 82501. This particular circuit replaces 12 integrated circuits required to perform the necessary interface functions with a single 20 pin IC. The circuit board can be reduced about 10 square inches due to the ESI 82501 IC.

The station-to-transceiver interconnect cable must be terminated in its characteristic 78 Ω differential mode and 18.5 Ω single mode. The ESI 82501 is equipped with the line receivers, decoder, encoder, and line driver necessary for the interface to communicate properly with the transceiver. Also included in the interface IC is the necessary signal-processing circuitry for collision detection and carrier presence. The interface has adequate interface circuitry to connect directly to the Intel 82586 controller as is shown in Figure 5–9(b). A multibus compatible board is available from Intel Corporation, which has all the necessary circuitry to support the controller and the interface unit.

In this section, the interface IC will be discussed rather than the Ethernet specifications, because the cost of these ICs will decrease and eventually they will be the common method of implementation. The controller will be discussed later in Chapter 6 which considers protocols. The controller has the protocol embedded within it. For the reader interested in details of the Ethernet specifications, see reference [4].

Functional Description

The functional description of the internal blocks is as follows:

Manchester Encoder and Transceiver Cable Driver

The 20 MHz clock is used to Manchester encode data on the TXD input line. The clock is also divided by 2 to produce the 10 MHz clock required by the 82586 for synchronizing its $\overline{\text{RTS}}$ and TXD signals. See Figure 3. (Note that the 82586 $\overline{\text{RTS}}$ is tied to the 82501 $\overline{\text{TEN}}$ input as shown in Figure 4.)

Data encoding and transmission beings with $\overline{\text{TEN}}$ going low. Since the first bit is a '1', the first transition on the transmit output TRMT is always negative. Transmission ends with the $\overline{\text{TEN}}$ going high. The last transition is always positive at TRMT and may occur at the center of the bit cell (last bit = 1) or at the boundary of the bit cell (last bit = 0). A one-bit delay is introduced by the 82501 between its TXD input and TRMT/$\overline{\text{TRMT}}$ output as shown in Figure 3. Following the last transition, the output $\overline{\text{TRMT}}$ is slowly brought to its high state so that zero differential voltage exists between TRMT and $\overline{\text{TRMT}}$. This will eliminate DC currents in the primary of the transceiver's coupling transformer. See Figure 4.

An internal watchdog timer is started at the beginning of the frame. The duration of the watchdog timer is 25 msec ±15%. If the transmission terminates (by deasserting the $\overline{\text{TEN}}$) before the timer expires, the timer is reset (and ready for the next transmission). If the timer expires before the transmission ends, the frame is aborted. This is accomplished by disabling the output driver for the TRMT/$\overline{\text{TRMT}}$ pair and deasserting $\overline{\text{CRS}}$. RXD and RXC are not affected. The watchdog timer is reset only when the $\overline{\text{TEN}}$ is deasserted.

The cable driver is a differential gate requiring external resistors or a current sink of 20 mA (on both terminals). In addition, high-voltage protection of 15 volts maximum for 1 second maximum is provided.

Receive Section

CABLE INTERFACE AND NOISE FILTER

The 82501 input circuits can be driven directly from the Ethernet transceiver cable receive pair. In this case the cable is terminated with a pair of 39-ohm resistors in series for proper impedance matching. The center tap of the termination is tied to an external voltage reference (source impedance of 18.5 ohms min) to establish the required common mode voltage bias for the 82501 receive circuitry. See Figure 4.

The input circuits can also be driven with ECL voltage levels. In either case, the input common mode voltage must be in the range of V_{CC} − 1.0 to V_{CC} − 2.5 volts to allow for a wide driver supply variation at the transceiver. The input terminals have a 15-

volt maximum protection and additional clamping of low-energy, high-voltage noise signals.

A noise filter is provided at the RCV/$\overline{\text{RCV}}$ input pair to prevent spurious signals from improperly triggering the receiver circuitry. The noise filter has the following characteristics:

A negative pulse which is narrower than 30 ns or is less than -150 mV in amplitude is rejected during idle.

At the beginning of a reception, the filter is activated by the first negative pulse which is more negative than -250 mV and is wider than 50 ns.

As soon as the first valid negative pulse is recognized by the noise filter, the $\overline{\text{CRS}}$ signal is asserted to inform the 82586 controller of the beginning of a transmission, and the $\overline{\text{RXC}}$ will be held low for 1.2 μsec maximum while the internal phase-locked-loop is acquiring lock.

The filter is deactivated if no negative transition occurs within 160 ns from the last positive transition.

Immediately after the end of a reception, the filter blocks all the signals for 5 μsec minimum, 7 μsec maximum. This dead time is required to block-off spurious transitions which may occur on the coaxial cable at the end of a transmission but are not filtered out by the transceiver.

MANCHESTER DECODER AND CLOCK RECOVERY

The filtered data enters the clock recovery and decoder circuits. An analog phase-locked-loop (PLL) technique is used to extract the received clock from the data, beginning from the third negative transition of the incoming data. The PLL will acquire lock within the first 12 bit times, as seen from the RCV/$\overline{\text{RCV}}$ inputs. During that period of time, the $\overline{\text{RXC}}$ is held low. Bit cell timing distortion which can be tolerated in the incoming signal is ±15 nsec for the preamble and ±20 nsec for data. The voltage-controlled oscillator (VCO) of the PLL corrects its frequency to match the incoming signal transitions. Its VCO cycle time stays within 5% of the RXD bit cell time regardless of the time distortion allowed at the RCV/$\overline{\text{RCV}}$ input. The RCV/$\overline{\text{RCV}}$ input is decoded from Manchester to NRZ and transferred synchronously with the receive clock to the 82586 controller.

At the end of a frame, the receive clock is used to detect the absence of RCV/$\overline{\text{RCV}}$ transitions and report it to the 82586 by deasserting $\overline{\text{CRS}}$ while $\overline{\text{RXD}}$ is held high.

Collision-Presence Section

The CLSN/$\overline{\text{CLSN}}$ input signal is a 10 MHz ±15% square wave generated by the transceiver whenever two or more data frames are superimposed on the coaxial cable. The maximum asymmetry in the CLSN/$\overline{\text{CLSN}}$ signal is 60/40% for low-to-high or high-to-low levels. This signal is filtered for noise rejection in the same manner as RCV/$\overline{\text{RCV}}$. The noise filter rejects signals which are less negative than -150 mV and narrower than 15 ns during idle. It turns on at the first negative pulse which is more negative than -250 mV and wider than 30 ns. After the initial turn-on, the filter remains active indicating that a valid collision signal is present, as long as the negative CLSN/$\overline{\text{CLSN}}$ signal pulses are more negative than -250 mV. The filter returns to the "off" state if the signal becomes less negative than -150 mV, or if no negative transition occurs within 160 ns from the last positive transition. Immediately after turn-off, the collision filter is ready to be reactivated.

The common mode voltage and external termination are identical to the RCV/$\overline{\text{RCV}}$ input. (See Figure 4.) The CLSN/$\overline{\text{CLSN}}$ input also has a 15-volt maximum protection and additional clamping against low-energy, high-voltage noise signals.

A valid collision-presence signal will assert the 82501 $\overline{\text{CDT}}$ output which can be directly tied to the $\overline{\text{CDT}}$ input of the 82586 controller.

During the time that valid collision-presence transitions are present on the CLSN/$\overline{\text{CLSN}}$ input, invalid data transitions will be present on the receive data pair due to the superposition of signals from two or more stations transmitting simultaneously. It is possible for RCV/$\overline{\text{RCV}}$ to lose transitions for a few bit times due to perfect cancellation of the signals. In any case, the invalid data will not cause any discontinuity of RXC.

When a valid collision-presence signal is present the $\overline{\text{CRS}}$ signal is asserted (along with $\overline{\text{CDT}}$). However, if this collision-presence signal arrives within 6.0 ± 1.0 μs from the time $\overline{\text{CRS}}$ was deasserted, only $\overline{\text{CDT}}$ is generated.

Internal Loopback

When asserted, $\overline{\text{LPBK}}$ causes the 82501 to route serial data from its TXD input, through its transmit logic (retiming and Manchester encoding), returning it through the receive logic (Manchester decoding and receive clock generation) to RXD output. The internal routing prevents the data from passing through the output drivers and onto the transmit output pair, TRMT/$\overline{\text{TRMT}}$. When in loopback mode, all of the transmit and receive circuits, including the noise filter, are tested except for the transceiver cable output driver and input receivers. Also, at the end of each frame transmitted in loopback mode, the 82501 generates a 1-μsec $\overline{\text{CDT}}$ signal within 1 μsec after the end of the frame. Thus, the collision circuits, including the noise filter, are also tested in loopback mode. The watchdog timer remains enabled in loopback mode, terminating test frames that exceed its time-out period.

In the normal mode ($\overline{\text{LPBK}}$ not asserted), the 82501 operates as a full duplex device, being able to transmit and receive simultaneously. This is similar to the external loopback mode of the 82586. Combining the internal and external loopback modes of the 82586 and the internal loopback and normal modes of the 82501, incremental testing of an 82586/82501-based interface can be performed under program control for systematic fault detection and fault isolation.

The pin description is given in Table 5–3.

The previous discussion has dealt with coaxial cable baseband hardware. The Intel ESI B2501 IC can also be used to implement the Institute of Electrical and Electronic Engineers (IEEE) standard 802.3. The next type of system to consider is the broadband coaxial cable system.

Broadband Cable Plants

The discussion will center on a cable plant similar to the Sytek bus. A description of a typical cable plant is given by Figure 5–10. The plant shown depicts the diversity of the data, which can be voice channels as indicated by the T1 carrier, terminal communication, and video distributed on a vestigial sideband carrier. The data are both analog and digital.

TABLE 5–3
Pin Description

Symbol	Pin No.	Type	Name and Function
$\overline{TXC}$	16	O	**Transmit Clock:** A 10-MHz clock output with 5 nsec rise and fall times. This clock is provided to the 82586 for serial transmission.
$\overline{TEN}$	15	I	**Transmit Enable:** An active low, TTL-level signal synchronous to $\overline{TXC}$ that enables data transmission to the transceiver cable. $\overline{TEN}$ can be driven by $\overline{RTS}$ from the 82586.
TXD	17	I	**Transmit Data:** A TTL-level input signal that is directly connected to the serial data ouput, TXD, of the 82586.
$\overline{RXC}$	8	O	**Receive Clock:** Clock output with 5 nsec rise and fall times and 50% duty cycle. This output is connected to the 82586 receive clock input $\overline{RXC}$. There is a maximum 1.4 μsec discontinuity at the beginning of a frame reception when the phase-locked loop switches from the on-chip oscillator to the incoming data. During idle (no incoming frames) the clock frequency will be half that of the 20 MHz crystal frequency.
$\overline{CRS}$	6	O	**Carrier Sense:** A TTL-level, active low output to notify the 82586 that there is activity on the coaxial cable. This signal is asserted when valid data or a collision signal from the transceiver is present. It is deasserted at the end of a frame synchronous with $\overline{RXC}$, or when the end of the collision-presence signal (CLSN and $\overline{CLSN}$) is detected, whichever occurs later.
RXD	9	O	**Receive Data:** An MOS-level output tied directly to the RXD input of the 82586 controller and sampled by the 82586 at the negative edge of $\overline{RXC}$. The bit stream received from the transceiver cable is Manchester decoded prior to being transferred to the controller. This output remains high during idle.
$\overline{CDT}$	7	O	**Collision Detect:** A TTL, active low signal which drives the $\overline{CDT}$ input of the 82586 controller. It is asserted as long as there is activity on the collision-presence pair (CLSN and $\overline{CLSN}$), and during SQE test in loopback.
$\overline{LPBK}/$ WDTD	3	I	**Loopback:** A TTL-level control signal to enable the loopback mode. In this mode, serial data on the TXD input is routed through the 82501 internal circuits and back to the RXD output without driving the TRMT/$\overline{TRMT}$ output pair to the transceiver cable. When $\overline{LPBK}$ is asserted, the collision circuit will also be turned on at the end of each transmission to simulate the collision test. The on-chip watchdog timer can be disabled by applying a 12V level through a 4k ohm resistor to this pin. $\overline{LPBK}$ must not be asserted at power up to ensure proper $\overline{CDT}$ and $\overline{CRS}$ signals to 82586 at start of operation.
TRMT	19	O	**Transmit Pair:** An output driver pair which generates the differential signal for the transmit pair of the Ethernet transceiver cable. Following the last transition, which is always positive at TRMT, the differential voltage is slowly reduced to zero volts. The output stream is Manchester encoded.
$\overline{TRMT}$	18	O	
RCV	4	I	**Receive Pair:** A differentially driven input pair which is tied to the receive pair of the Ethernet transceiver cable. The first transition on RCV will be negative-going to indicate the beginning of a frame. The last transition should be positive-going, indicating the end of a frame. The received bit stream is assumed to be Manchester encoded.
$\overline{RCV}$	5	I	
CLSN	12	I	**Collision Pair:** A differentially driven input pair tied to the collision-presence pair of the Ethernet transceiver cable. The collision-presence signal is a 10 MHz ±15% square wave. The first transition at CLSN is negative-going to indicate the beginning of the signal; the last transition is positive-going to indicate the end of the signal.
$\overline{CLSN}$	11	I	
C1	1	I	**PLL Capacitor:** Phase-locked-loop capacitor inputs.
C2	2	I	
X_1	14	I	**Clock Crystal:** 20-MHz crystal inputs.
X_2	13	I	
V_{CC}	20		**Power:** 5 ± 10% volts.
GND	10		**Ground:** Reference.

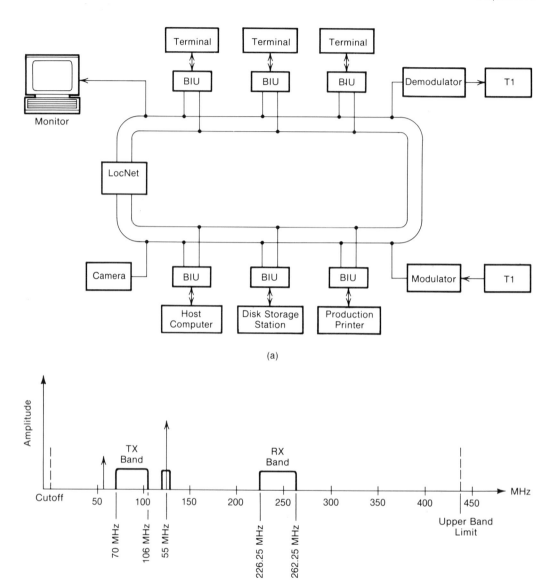

Figure 5-10 (a) A Sytek broadband cable plant. (b) Frequency spectrum of the channels on the broadband cable.

The video data must modulate a carrier; baseband video cannot be used, because the cable amplifiers are not dc coupled. Also, only one video channel would be available at baseband. The cable bandwidth is 5 MHz to 300 or 400 MHz, depending on the cable used, which can be CATV cable.

A spectral plot of the cable is shown in Figure 5-10(b). Note that a great deal of the spectrum is not used even with all the implements shown. Other implements such as facsimile, high-speed modems for faster data transmission, and gateway stations to foreign LANs may be added later.

The bus interface units (BIUs) have frequency agile RF modems that will tune to one of the 20 logical channels of the Sytek LocNet 20. The BIUs each have a group bandwidth within the transmit (70 to 106 MHz) and receive (226.25 to 262.25 MHz) band depending on their group letter. Groups have a 6-MHz bandwidth, with each FDM channel within the group having a bandwidth of 300 kHz. The transmit and receive bands are both 36 MHz wide, which implies that

$$\text{Maximum no. channels} = 6 \text{ groups} \times 20 \text{ channels/group} = 120 \text{ channels}$$

The BIU will transmit or receive data from the broadband cable at 128 KBits/s using the frequency agile modem to encode the data frequency-shift keying (FSK) or to demodulate incoming data. A higher-speed transmission is possible using LocNet 40, which has five high speed 2MBit/s channels on this same cable. Bridges may be used to extend 20 channel groups to other groups. This will provide a single integrated network.

Each BIU is connected to a directional coupler, which is a coaxial tap. A small amount of energy is tapped from the cable as determined by the tap ratio. These taps disturb the line very little, which eliminates the need for tuning or matching.

As an example of how directional couplers operate, if a signal on the cable is 34 dBmV at the tap, where dBmV is defined as

$$y \text{ dBmV} = 10 \log \frac{x}{1 \text{ mV}} \qquad (5\text{--}20)$$

and the BIU requires an input signal of -2 dBmV $\pm$ 6 mV, the nominal tap ratio is 34 dBmV $-$ (-2 dBmV) $=$ 36 dB. A 35.1-dB tap is available commercially with a signal level of -1.1 dBmV with this coupler. The transmitter power is adjustable for this BIU, but it may not be for others. A tap ratio will be needed to prevent overdrive of the amplifiers. This same technique must be used for other signals as well. Interface circuits for other forms of data where BIUs are not used may require external amplification at the directional coupler output. The reason for keeping signals low at BIU and coaxial cable amplifiers is to prevent intermodulation products from occurring.

Note that the headend has a Sytek 50/50. This unit translates the signals in the transmit band to the receive band (i.e., 70 to 106-MHz signals to 226.25 to 262.25 MHz). When a station begins transmitting, the receiver at that station will receive a delayed version of its data. The receiver will compare the source address of the received data; source and data can be checked for errors. Some of the advantages will be explored when protocols are analyzed in Chapter 6.

The modems within the BIUs are rather complex because they are frequency

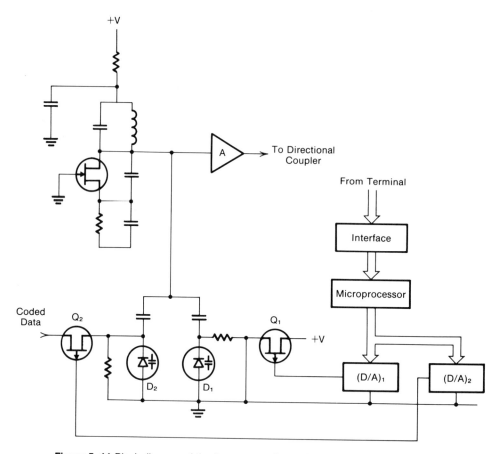

Figure 5-11 Block diagram of the frequency agile modem.

agile and are tuned digitally. An example of such a modem is shown in block diagram form in Figure 5-11.

The theory of operation is as follows: Varacap or hyperabrupt tuning diodes (capacity is a function of voltage) are used to tune the oscillator. The tuning element D_1 will tune the Colpitts oscillator to the channel frequency. The channel number, if entered in the microprocessor, is converted to a binary number, which sets the digital-to-analog converter (D/A) voltage output, in conjunction with the FET Q_1, to bias D_1 correctly to set the channel frequency. Calibration curve hyperabrupt tuning diodes (capacity is a function of voltage) are used to tune the oscillator. Calibration curves could be stored in the microprocessor memory. Diode D_2 will frequency modulate the oscillator with the deviation set by Q_2. A calibration curve is also needed to correct $(D/A)_2$ settings.

The modem shown in Figure 5-11 is oversimplified and rather expensive. It does not represent the Sytek modems. At the present time, a great deal of research is being conducted on high-frequency modems, and the manufacturers of these devices are very reluctant to allow the circuitry to be published. But, as one may observe even with this trivial example, the circuitry is complex. When constructing a broadband network, the reader will soon discover it much less costly to buy these components rather than design and fabricate them. The circuitry for other components in the network is readily available should the reader wish to delve into them further. General Instrument is a good source of material for CATV components such as directional couplers, wideband amplifiers, and filters. T1 Specification information is available from Bell Telephone or Western Electric. More discussion on broadband networks is reserved for Chapter 7.

Fiber-optic Cable Plants

The final cable plant for discussion is a fiber-optic plant. Fiber optics have a particular property, which makes them rather appealing; as the bandwidth of transmission increases, core sizes will generally be decreased. The opposite is true for coaxial transmission line (i.e., as bandwidth increases, cable size will also increase). At the end of this section, the advantages and disadvantages of fiber optics will be explored.

A cable plant is shown in Figure 5–12 with all the elements that cause loss. This is not intended to be a typical plant. A typical plant, for example, may be implemented with a fiber-optic switch and no star, or vice versa. Only an optical path in from transmitter to receiver is shown. Typical configurations depend on the topology (i.e., ring, star, bus, etc.). The components will now be examined. Only a functional description will be given for each.

The transmitter consists of electronics for encoding the binary signals for a baseband transmitter and drive components for semiconductor light-emitting diode (LED) or laser sources. If the transmitter is an analog type, it accepts an analog signal with some form of modulation (AM, FM, PM, FSK, etc.) and this signal will in turn intensity modulate a LED or laser source. Laser sources produce signal optical power levels between 1 and 10 mW (0 to 10 dBm), and LEDs produce optical power levels from 100 to 1000 μW (-20 to 0 dBm).

The Tee coupler may be used for either summing two optical signals or as a tap. The summation of signals may be of two different wavelengths or two different frequencies intensity modulating the same wavelength. The Tee coupler can be used as a tap and will operate similar to a directional coupler (i.e., the Tee may have large tap ratio to take only a small amount of optical power from the main path).

The star coupler sums all the input signals and the power is distributed among its output ports. The summation may consist of multiple wavelengths or identical wavelengths with different frequencies intensity modulating each on the same wavelength. The summation is shown in Equation 5–21. Each input must

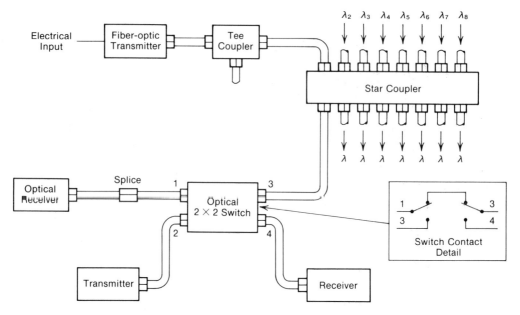

Figure 5-12 Fiber-optic components in the transmit-to-receive path.

be distributed between eight outputs and the loss (excess loss) of the star must also be considered. Stars have losses as large as 16 to 18 dB, which is shown in Equation 5–22.

$$P_{in} = \text{input power} = \sum_{n=1}^{\infty} P_{\lambda ni} \qquad (5\text{-}21)$$

$$P_{out} = \text{output power} = \frac{1}{n}\sum_{n=1}^{k} P_{\lambda no} - \text{losses} \qquad (5\text{-}22)$$

The switch operates similarly to a double-pole double-throw set of contacts configured as shown in the switch contact detail. It functions as a bypass in this configuration for a transmitter and receiver. If this were a node in a ring topology and the node had power failure or damage, this switch would allow the node to be bypassed. There is a variety of these switches on the market with different contact configurations. The loss of these devices in the bypass condition is generally from 1 to 2 dB.

Splices are used to extend the length of the cable in the cable plant beyond 2 km. Fiber-optic cables can be purchased in 1- and 2-km reels. Splices are sometimes used in place of connectors because they have lower loss (0.1 to 0.05 dB). For example, when using a star the loss is large, and splices may be substituted for connectors to improve the loss budget.

Physical Layer

The next item to consider is the connectors. Connectors produce between 0.5 and 1 dB of loss. As one can observe in the diagram, the path has eight connectors from transmitter to receiver, which would represent a substantial loss. Connectors and connector technology can cover volumes of material; therefore, only the approximate loss will be considered here.

The last item to be considered is the receiver. For our discussion, the receiver will consist of a detector that is a PIN or APD diode, preamplifier, output amplifier (analog or digital), decoder for baseband digital, and automatic gain control circuit. For the digital receiver, the decoded output will be used with binary waveforms, and for the analog case some form of modulated carrier will be present. The topics have been discussed superficially, but the reader who wishes to delve into the subject further may consult reference [5].

The next topic to be considered is the loss budget. This is most important because it will affect optical design and system performance. The losses in the optical part of the system are added up, and an assessment of the receiver sensitivity and transmitter output power needed to get the necessary performance can be determined.

Depending on the loss budget, one determines whether to use an LED or laser transmitter. Also, the detector may be selected, either a PIN or APD type. After a loss budget is generated, the designer can examine it for the trade-offs that will make the design more cost effective. A loss budget check-off list is shown in Table 5–4.

As an aid to the reader, an explanation of each variable is given with typical values.

X_1	Transmitter output power is self-explanatory; typical range, LED (-20 to 0 dBm) and lasers (up to 10 dBm in solid state).
X_2	This is the coupling loss; only 0.1 to 1.5 dB is typical.
X_3	-2.5 to -8 dB/km, 50 nm at $\lambda = 820$ nm is typical.
X_4	Range of LAN application: ½ to 10 km.
X_5	Calculated: $X_5 = X_3 \times X_4$.
X_6	This is the number of switches allowed in the bypass condition between the transmitter and receiver.
X_7	Insertion loss per switch is self-explanatory (1 to 2 dB is typical).
X_8	Total switch insertion loss $= X_6 \times X_7$.
X_9	Switch termination losses. Self-loss $= X_6 \times X_7$.
X_9	Switch termination losses. Self-explanatory from the table.
X_{10}	Self-explanatory.
X_{11}	Self-explanatory, with typical values of 0.05 to 0.1 dB.
X_{12}	Total loss due to splices $= X_{10} \times X_{11}$.

TABLE 5-4
Loss Budget

X_1	Transmitter output power _____dBm	L_T
X_2	Transmitter termination loss _____dB	L_C
X_3	Cable attenuation _____dB/km	
X_4	Cable length _____km	
X_5	Cable attenuation _____dB	L_{ATT}
X_6	Number of fiber-optic switches _____	N_{SW}
X_7	Insertion loss/switch _____dB	L_{IN}
X_8	Total switch insertion loss _____dB	
X_9	Switch termination: $2N_{SW} \times L_C$ _____dB	L_{SC}
X_{10}	Number of splices _____dB	N_{SP}
X_{11}	Splice loss _____dB	L_{SP}
X_{12}	Total splice loss _____dB	L_{TSP}
X_{13}	Star-coupler termination loss _____dB	L_C
X_{14}	Insertion loss, star coupler _____dB	
X_{15}	Output power in dBm star coupler _____dBm	
X_{16}	Receiver termination loss _____dB	
X_{17}	Receiver minimum sensitivity _____dBm	
X_{18}	Temperature variation allowance _____dB	
X_{19}	Format-NRZ, -3 dBm; RZ, -6 dBm	
X_{20}	Aging loss _____dB	L_a
X_{21}	Transmitter power end of life _____dBm	
X_{22}	Required SNR _____dB or BER _____	
X_{23}	Source output halved increase lifetime -3 dBm	
X_{24}	Miscellaneous losses _____dB	
X_{25}	Total losses coupling insertion, etc. _____dB	
X_{26}	Adjusted power output _____dBm	
X_{27}	Total available power margin _____dB	
X_{28}	Power margin _____dB	

X_{13} Input and output termination loss for the star coupler same as other termination.

X_{14} Insertion loss may be given by manufacturer or calculated for the typical range of 1 to 3 dB.

X_{15} It is calculated or given by the manufacturer and is application dependent.

X_{16} Self-explanatory; range of values typically 0.5 to 1.5 dB.

X_{17} This figure is given by the manufacturer or must be calculated during receiver design (-30 to -50 dBm is typical).

X_{18} This parameter deals with all temperature-related losses after the transmitter and ahead of the receiver (such as switch, splice, cable, couplers, connectors, etc.). The transmitter and receiver variations are taken into account when specifying transmitter output and receiver sensitivity. Variations may occur in the alignment of the components of the link, which will cause losses.

X_{19} Self-explanatory.

X_{20} Aging loss again considers only fiber-optic components after the transmitter and ahead of the receiver. This term deals with cable deterioration, splices, switches, and the like.

X_{21} Transmitter power will diminish with age. Therefore, when the power output has dropped to a specific percentage of the original minimum value, end of life is reached and the transmitter should be replaced (usually 80 percent of the original).

X_{22} The SNR or BER must be given to determine if performance can be maintained over the lifetime of the network. If BER is given, the SNR required can be determined from Table 5-5 or calculated.

TABLE 5-5
Bit Error Rate Versus Signal-to-Noise Ratio

BER	SNR	BER	SNR	BER	SNR
10^{-2}	13.5	10^{-5}	18.7	10^{-9}	21.6
10^{-3}	16.0	10^{-6}	19.6	10^{-10}	22.0
10^{-4}	17.5	10^{-7}	20.3	10^{-11}	22.2
		10^{-8}	21.0		

X_{23} Operating LEDs or lasers (in particular) at half their rating will increase device life. However, many new devices have sufficiently long mean time between failures that this is no longer required. But, for installations to be used for greater than 5 or 6 years, such as the telephone industry, this may need to be considered.

X_{24} Miscellaneous items deal with all the items not accounted for in other terms in the budget such as receiver pigtail and cable waveguide or mismatches. These mismatches can occur on items such as switches, star couplers, and Tee couplers.

X_{25} Total loss must be summed as shown by Equation 5-23.

$$X_{25} = X_2 + X_5 + X_8 + X_9 + X_{12} + X_{13} \\ + X_{14} + X_{16} + X_{18} + X_{20} + X_{24} \quad (5\text{-}23)$$

X_{26} This is the power output available after all adjustments and deratings are made as calculated using Equation 5-24.

$$X_{26} = X_{19} + X_{21} + X_{23} + X_1 \tag{5-24}$$

X_{27} This is the total amount of power margin available before any link losses are considered. The calculation is made using Equation 5–25.

$$X_{27} = X_{26} - X_{17} \tag{5-25}$$

X_{28} The excess power margin is calculated with Equation 5–26. If the value is negative, this indicates the link will only marginally operate, if at all.

$$X_{28} = X_{27} - X_{25} \tag{5-26}$$

A comparison must be made of the excess power margin with the required SNR with the restrictions shown in the calculation of Equation 5–27.

$$X_{28} - X_{22} \geq 0, \quad \text{for adequate performance} \tag{5-27}$$

An example of a fiber-optic ring will be examined to illustrate the use of Table 5–4.

Example 5–1

A LAN is to be designed with the following specifications:

Maximum terminal distance	2 km
Cable attenuation	4 dB/km
Connector loss	1 dB/termination
Switch insertion loss	2 dB/switch
Maximum number of switch bypasses between terminals	2
Maximum number of splices	3
Loss/splice	0.5 dB
Launch power, 100 μW	−10 dBm
Code format	NRZ
Receiver worst-case sensitivity	−42 dBm
BER at 10^{-9} worst-case SNR	10 dB
Power ratio	$\dfrac{100 \text{ Nw}}{1 \text{ mw}}$

A table of values has been computed for the specifications given. Loss budget for the example is also presented, with values labeled N/A meaning "does not apply." The value is calculated to indicate power margins: If the BER/SNR subtraction is negative, an adjustment must be made to the link. General trade-offs can be made as follows:

1. The transmitter power can be increased.
2. A more sensitive receiver can be substituted.
3. Lower-loss cable can be used in the system.
4. Lower-loss connectors can be used.
5. Fewer bypass switches can be allowed.
6. Lower-loss switches can be used.
7. All switches can be purchased with pigtails and spliced into the fiber-optic circuit. This technique will make the terminals less portable.

Loss Budget for the Example

Transmitter output power -10 dBm
Transmitter termination loss $\underline{1}$ dB
Cable attenuation $\underline{4}$ dB/km
Cable length maximum $\underline{2}$ km
Cable attenuation $\underline{8}$ dB
Number of fiber-optic switches $\underline{2}$
Insertion loss/switch $\underline{1.5}$ dB
Total switch insertion loss $\underline{3}$ dB
Switch termination loss, $2N_{SW}XL_C\underline{4}$ dB
Maximum number of splices $\underline{3}$
Splice loss $\underline{0.5}$ dB
Total splice loss $\underline{1.5}$ dB
Star couplers' termination loss $\underline{N/A}$ dB
Output power of star coupler $\underline{N/A}$ dB
Insertion loss, star coupler $\underline{N/A}$ dB
Receiver termination loss $\underline{1}$ dB
Receiver minimum sensitivity -42 dBm
Temperature variation allowance $\underline{1}$ dB
Format NRZ -3 dBm
Aging loss $\underline{1}$ dB
Transmitter power at end of life (assume 80% of worst-case) -1 dB
Source output not halved
Miscellaneous loss $\underline{N/A}$ dB
Equation 5–23: $X_{25} = 1 + 8 + 3 + 4 + 1.5 + 1 + 1$
$\qquad\qquad\qquad = 19.5$ dB
Total loss 19.5 dB
Equation 5–24: $X_{26} = (-3 \text{ dBm}) + (-1 \text{ dBm}) + (0) + (-10 \text{ dBm})$
Adjusted power output $\underline{14}$ dBm
Equation 5–25: $X_{27} = (-14 \text{ dBm}) - (-42 \text{ dBm})$
Total available power margin $\underline{28}$ dB
Equation 5–26: $X_{28} = 28 - 19.5$

Power margin <u>8.5</u> dB (actual margin if BER is included in sensitivity specification, which is the usual case for digital receivers)

$$8.5 - 10 \text{ dB} = -1.5$$

When configuring a terminal, remember to keep the fiber-optic terminations at a minimum. As an example, connections from the transmitter and receiver to the fiber-optic switch should be made with splices. The switch should be acquired with sufficiently long pigtails to allow connections to be made with bulkhead connector through the terminal enclosure wall. These precautions will keep losses minimized within the machine enclosure. The specification did not include all parameters; these will allow the reader to focus on one of the series of calculations at a time.

The next parameter to consider is rise time. A table of calculations similar to the loss budget can be generated, as follows:

Rise Time Calculations

T_1 Total system rise time _____ns
  Digital RZ, 0.35/baud rate
  Digital NRZ, 0.7/baud rate
  Intensity modulation analog, 0.35/bandwidth

$$\text{Pulse position modulation (PPM)} = \left[\text{Sampling rate} \times \left(\frac{\text{Pulse separation}}{\text{pulse width}} \right) \times \text{bandwidth} \right]^{-1}$$

$$\text{Pulse code modulation (PCM)} = \left[\text{Sampling rate} \times \left(\frac{\text{BITS}}{\text{SMPL}} \right) \times \text{bandwidth} \right]^{-1}$$

T_{2a} Transmitter rise time, analog
T_{2b} Transmitter rise time, digital _____ns
T_3 Detector rise time _____ns
  Waveguide NA _____, N _____
  $\lambda = $ _____nm, λ_S _____nm
T_4 Modal dispersion approximation of rise time
  Step index $T_s/l = 600 N [1 - \sqrt{(1 - NA)^2}]$
  Graded index, $TG/l = T_{S/20}$
  Loop/ring $T_4 = (TG/l \text{ or } TS/l) \times \Sigma \text{ length}$ _____ ns
  Star $T_4 = (TG/l \text{ or } TS/l) \times L_{MAX}$ _____ ns
T_5 Material dispersion approximation of rise time
  $T_{MD/l} = \lambda \times \lambda_S \times 10^{-4}$ (step or graded index)
  Loop/ring $T_5 = T_{MD/l} \times \Sigma \text{ length}$ _____ ns
  Star $T_5 = T_{MD/l} \times L_{MAX}$ _____ ns
T_{total} = total rise time

$$T_{total} = 1.11 \sqrt{T_2^2(T_{2a} \text{ or } T_{2b}) + T_4^2 + T_5^2}$$

If added to the previous specification, the following rise time parameters are generated.

Rise Time Calculation Example 5–2

$\lambda = 820$ nm, $\lambda_s = 20$ nm, graded index
Baud rate, 6 Mbits/s
$NA = 0.20$, $N_{CLAD} = 1.5$
Digital detector rise time, 3.0 ns
Transmitter rise time, 5.0 ns
PCM is the modulation technique
T_1 Total system rise time $\underline{8.1}$ ns

$$\text{PCM} = [2.5 \text{ sampling rate} \times (40 \text{ dB}/6 + 1.2) \times 6 \times 10^6]^{-1}$$

T_{2b} Transmitter rise time $\underline{5}$ ns
T_3 Detector rise time $\underline{2.5}$ ns
T_4 Modal dispersion rise time $\underline{1.21}$ ns

$$l = 2 \text{ km}, \; T_{G/l} = (600/20)[1 - \sqrt{1 - 0.2^2}]$$

T_5 Material dispersion rate time $\underline{3.28}$

$$T_{MD/l} = 820 \times 20 \times 10^{-4}, \; T_s = \frac{T_{MD}}{l} \times 2 \text{ km}$$

$T_{\text{TOTAL}} = 1.11 \sqrt{T_2^2 + T_3^2 + T_4^2 + T_5^2}$
$T_{\text{TOTAL}} = 7.31$ ns, $T_{\text{TOTAL}} < T_1$

Comparing system rise time T_1, the required rise time, and that of the calculation T_{TOTAL} indicates it is adequate to meet system requirements. The transmitter and receive rise time shown are the end result after taking all amplifiers and drive circuits into account. The receiver and transmitter rise times can be measured or calculated values. If these devices are purchased as modules, the specification will include rise time. If design of these components is part of the effort, then the transmitter and receiver values are calculated or measured.

Bandwidth calculations must be made to determine if the system is bandwidth limited. If rise time is adequate, then the bandwidth will allow correct system performance. Occasionally, bandwidth figures are only known for transmitter and receiver. The equations for calculating the rise time can be employed because of bandwidth dependence. The cable is specified in MHz-km, but it is usually only given at a particular wavelength. If a wavelength other than specified is used, the overall rise time can be calculated using the preceding format, and the bandwidth can be calculated using one of the equations in the T_1 entry, depending on the modulation technique. For our previous example, the system rise time was 7.31 ns. Then

$$T_R = (0.7/B_{\text{rate}}) \quad \text{for digital NRZ}$$
$$B_{\text{RATE}} = 95.5 \text{ Mbaud} \quad \text{for NRZ}$$

Fiber-optic Transmitters and Receivers

There are large numbers of combinations, but due to space limitations the scope of the investigation will be limited to a digital LED transmitter, laser analog transmitter, PIN analog receiver, and APD digital receiver.

A schematic is shown in Figure 5–13 of a high-performance LED transmitter. Simpler transmitters may be built with a logic gate driving a transistor, with the LED located in the emitter or collector circuit. The circuit shown is superior to the previous two circuits mentioned. Switching large currents will cause a great deal of power line noise; the circuit in the figure will sacrifice power consumption for reduced noise on power lines and increased switching speed. Note that an extra logic gate is added to the circuitry. This configuration will add a little extra delay to the circuitry, but it corrects for the differences in rise and fall times. The gate will reduce duty-cycle distortion to zero. Decoupling added to the power supply inputs is an extra precaution to minimize noise from the transmitter. Receivers should not be placed in close proximity to transmitters due their high gains (3000). Other circuits for higher-performance transmitters may be found in reference [5].

An analog transmitter is shown in block diagram form in Figure 5–14. This circuit has two types of control circuitry as indicated in the diagram. The bias-control circuit is composed of optical feedback, which modifies the bias current of the laser to correct for changes in its optical power output. Power output is affected a great deal by temperature excursions during transmission. Thus the

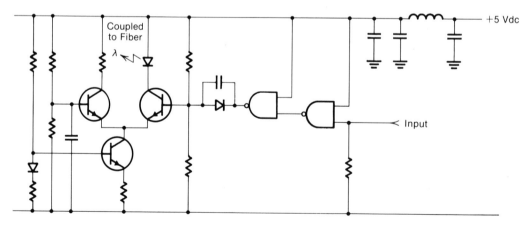

Figure 5–13 Fiber-optic digital transmitter schematic.

addition of a temperature-control circuit offers even tighter control over the operating point.

When the laser is operating with no RF input, current flows through D_2, L_1, R_5, Q_1 and R_{10} and produces a continuous optical power level at the fiber-optic pigtail. As RF power is injected into the circuit path R_1, C_1, R_2, R_3, C_2, and L_2, the laser is intensity modulated with the signal. The inductance L_1 and capacitor C_3 decouple the RF from the dc bias circuit. When the power level of the laser increases, photodetector D_3 senses the increase and a signal is fed back to amplifier A_1. The feedback at the input of A_1 reduces the base bias voltage at Q_1; this

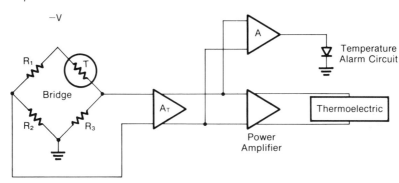

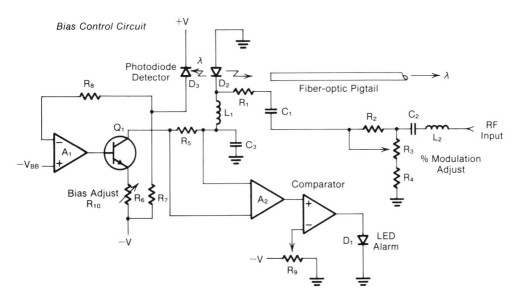

Figure 5–14 Laser analog transmitter block diagram with temperature control and alarm circuits.

reduces the bias current to the laser because Q_1 begins to turn off. If the analysis is followed in a similar manner for the reverse situation, the optical output power level has decreased. The following events occur to correct the bias current: Detector output is reduced, thus decreasing the signal input to A_1; this increases the base voltage on Q_1, turning it on further. As a result of Q_1 turning on, more current flows through the laser.

The amplifier A_2, comparator, R_5, R_9, and D_1 form an alarm circuit. Resistor R_5 is 1 ohm (Ω); therefore, the output of amplifier A_2 can be made proportional to current if A_2 has a gain of 1. The comparator is adjusted to turn on LED D_1 when the current through the laser reaches a critical value.

Lasers are very sensitive to being overdriven. Generally, they will degrade slowly (i.e., power output will be reduced as facets sustain more damage). Occasionally, the laser will appear to be normal but, after the transmitter is turned off and allowed to cool, the laser will produce very low power when turned on again.

To prevent laser degradation due to temperature and to maintain a stable operating point for bias current, a temperature-control circuit is added. The thermistor for sensing temperature is in the bridge circuit as shown. As the temperature of the thermistor changes, amplifier A_t will amplify the difference, which will drive the power amplifier. The power amplifier will apply voltage across the thermoelectric device, which will cause heating or cooling of the laser. The temperature-control circuit has a much longer time constant than the bias circuit. This situation will cause a slowly varying modulation of the laser output, which is dependent on the thermal mass.

These two transmitters were presented with simplified circuits to illustrate some of the techniques used for transmitter design. Commercially available transmitters are preferred because they offer the designer a compact package with compensation built into the device. Often the transmitter can be purchased with the encoders also built in, which will allow both clock and data to be superimposed on the same waveform (the last three waveforms in Figure 5–1). These techniques will be addressed later in this chapter. Two of the most important parameters to be considered are transmitter output power and SNR or BER. Power output must be sufficient to overcome all losses in the system and produce a signal that can be detected and amplified undistorted by the receiver. The signal must meet some criteria to determine whether it is usable, such as SNR for the analog case and BER for digital systems.

Most of the laser transmitters have SNRs of 50 to 60 dB. This may seem very large, but as stated previously video distribution systems using standard broadcast modulating techniques will severely limit the end-to-end link loss. LEDs have intrinsic noise, which is difficult to measure; therefore, they will not be considered here.

To make the discussion complete, receivers must be considered. The two types presented (PIN and APD) are popular for use in fiber-optic technology. The receiver shown in Figure 5–15(a) is a PIN analog type. If the receiver is equipped

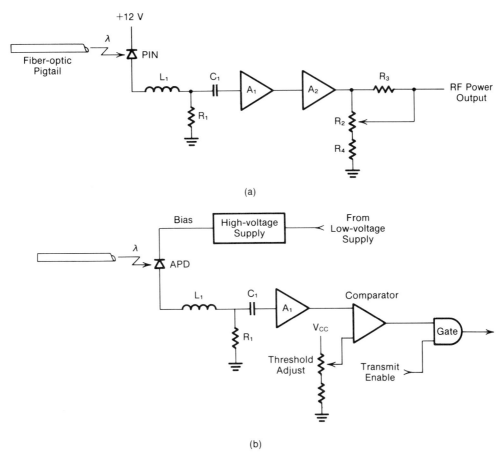

Figure 5–15 (a) Simplified PiN receiver. (b) Simplified digital APD receiver.

with a zero crossing detector at the output of A_1 (preamplifier) and high-frequency roll-off is added, this type of receiver may be used for digital signal recovery. Also, if the dynamic range is large, some form of automatic gain control (AGC) will be necessary. The PIN detector will convert incoming signals from optical to electrical, with L_1 used to extend the receiver bandwidth. This particular receiver is ac-coupled to A_1, the transimpedance preamplifier. These preamplifiers are useful because they convert the detector (current source) to a voltage signal, which is more useful as an output. A_2 produces an analog output voltage, with R_3 producing a level adjustment.

Equation 5–28 can be used to calculate the RF output signal.

$$E_{avg} = \frac{P_{avg} m \gamma'}{\sqrt{2}} \qquad (5\text{--}28)$$

where P_{avg} = average optical power input
m = modulation index
γ' = responsivity
  = $\gamma A_1 A_2$
  = $\gamma \dfrac{\mu a}{\mu W} \times \beta \dfrac{mV}{\mu A} \times \gamma$

The units of γ are microamperes per microwatt ($\mu A/\mu W$) or amperes per watt:

$$E_0 = \frac{P_{avg} m \gamma A_1 A_2}{\sqrt{2}} \qquad (5\text{-}29)$$

For a typical application of Equation 5-29, the parameters have the following values:

Example 5-2

$P_{avg} = 100\ \mu W, \quad m = 0.5\ (50\%\ \text{modulation})$

$\gamma = 0.5\ \dfrac{\mu A}{\mu W}, \quad A_1 = 10,\ A_2 = 1000$

$E_0 = 176.7\ mV \quad \text{or} \quad 45\ dBmV$

The bandwidth for a typical receiver of this type can be 200 to 300 MHz.

The next receiver to consider is the APD type, which will be examined for the digital case only. This does not imply that these receivers are useful for digital circuitry only. The advantages of APD are the following:

1. They have low dark current, which appears in the SNR as a noise component.
2. Better quantum efficiency.
3. Low noise.

There are trade-offs that will allow increased quantum efficiency for bandwidth, depending on the construction.

A block diagram of a typical APD digital receiver is shown in Figure 5-15(b). The block depicting a high-voltage supply is necessary for back-biasing the APD; this voltage may be 100 to 300 V. The practical gain limitation of these detectors is 40 to 100. APDs are gain-bandwidth product limited; therefore, a trade-off can be made: gain for bandwidth, or vice versa.

Equation 5-29 can be modified to reflect the addition of G_a, the avalanche gain, with the result given by Equation 5-30.

$$E_0 = \frac{P_{avg} m \gamma G_a A_1 A_2}{\sqrt{2}} \qquad (5\text{-}30)$$

The calculations made using Equations 5–29 and 5–30 ignore noise in the system.

One important item that must be considered in receiver design is SNR. For the coaxial and twisted shield pair LANs, this topic was ignored or treated lightly because many sources are available that address it (for example, see reference [1]). For the fiber-optic case, some treatment of the subject will be given here. Equation 5–31 represents the SNR of an amplifier with a PIN detector.

$$\text{SNR} = \frac{(\gamma P_0 m)^2}{2\gamma e P_0 B_n + N_T^2 + N_d^2}$$

where $N_T^2 = \dfrac{4KTB_n}{R_L}$, thermal noise component

$N_d^2 = 2eI_dB_n$, dark current noise

$\gamma = \dfrac{\eta e}{h\nu}$, responsivity (optical variable)

$2\gamma e P_0 B_n$ = optically generated shot noise

Often the noise equivalent power (NEP) is desired. This is when the signal power equals noise power. It is given by Equation 5–32.

$$\text{SNR} = \frac{(\gamma P_0 m)^2}{2\gamma e m P_0 B_n + 2eI_d B_n + \dfrac{4KTB_n}{R_L}} \qquad (5\text{-}31)$$

Note that several variables must be specified if the NEP is to be valid; therefore, NEP values are meaningless unless all the conditions of measurement are specified.

$$\gamma^2 P_0^2 m^2 = 2\gamma e P_0 B_n + 2eI_d B_n + \frac{4\,KTB_n}{R_L}$$

$$P_0^2 = \frac{2\,eP_0 B_n}{\gamma m^2} - \frac{2\,eI_d B_n}{\gamma^2 m^2} - \frac{4\,KTB_n}{\gamma^2 m^2 R_L} = 0$$

Then

$$P_0 = \frac{eB_n}{\gamma m^2} \pm \frac{1}{2}\sqrt{\frac{e^2 B_n^2}{\gamma^2 m^2} + 4\left(\frac{2eI_d B_n}{\gamma^2 m^2} + \frac{4KTB_n}{\gamma^2 m^2 R_L}\right)}$$

$$\text{NEP} = \frac{eB_n}{\gamma m^2} \pm \frac{1}{2\gamma m}\sqrt{\frac{e^2 B_n^2}{m^2} + 8eI_d B_n + \frac{16KTB_n}{R_L}} \qquad (5\text{-}32)$$

The normalized version if no optical detection were present would be in power per root hertz.

If, for example, the detector in the previous PIN receiver has a small dark current noise contribution, the SNR is calculated as

$$SNR = \frac{(0.5\ \mu A/\mu W \times 100\ \mu W \times 0.5)^2}{2 \times 0.5\ \frac{\mu A}{\mu W} \times 1.6 \times 10^{-19} C \times 100\ \mu W \times 300\ MHz + \frac{4.8 \times 10^{-12}}{10^3}}$$

$= 34.16$ dB, at a noise bandwidth of 300 MHz with a detector load of $R_L = 1\ k\Omega$.

As one can observe, the SNR would cause a poor video display, as discussed previously; but if the signal were digital baseband, the SNR would be much larger than needed to produce BERs of less than 10^{-9}.

The SNR equation for an APD detector and receiver is given in Equation 5–33.

$$SNR = \frac{(\gamma P_0 Gm)^2}{2\gamma e P_0 m G^2 F(G) B_n + 2e I_d B_n + \frac{4KTB_n}{R_L}} \qquad (5\text{–}33)$$

where

$$F(G) = G\left[1 - (1-K)\frac{(G-1)^2}{G^2}\right]$$

$K = 0.03$ to 0.1 for silicon diodes
$K \approx 1$ for germanium diodes
$K =$ hole-to-electron ionization ratio

High-frequency models of the analysis may be found in reference [6] with the derivations of the design equations.

Comparisons Between Fiber Optics and Copper Cable Technology

A comparison of fiber-optic and copper-cable technology will be presented to give the reader some insight into performance limitations. Consequently, neither technology is a solution for all network design problems.

One of the most obvious characteristics of fiber-optic transmission is the low loss and large bandwidth of the medium (see Figure 5–16). As can be observed from the curves, single-mode fiber shows a great deal of promise for long-distance transmission. Repeaterless transmission for fiber-optic single-mode fibers has been recorded at distances over 140 km. Commonly designed links can easily accommodate repeaterless transmission for 70 km, while for coaxial links maximum repeaterless transmission is limited to approximately 50 km. Multimode links will operate at about the same distance when repeaterless.

Crosstalk and interference between adjacent signal paths will cause problems

Physical Layer

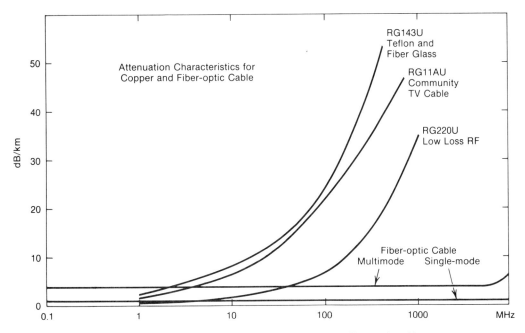

Figure 5–16 Transmission characteristics of copper and fiber-optic cable.

in copper cables, whereas fiber-optic cables are nonconductive and are not affected by extraneous signals and noise. Also, fiber-optic cables do not radiate and produce interference.

Fiber-optic cable, when compared to copper lines, will carry a great deal more traffic than copper cable. As an example, a 3½-in. duct with 320 twisted pairs or 20 coaxial cables has the same information-carrying capacity as a ⅜-in.-diameter five-fiber optical cable.

Four important advantages of fiber-optic transmission for military use are a high measure of security, high immunity to damage from electromagnetic pulse (EMP) after a nuclear explosion, light weight, and small size. Optical cable can be tapped, but with great difficulty, and countermeasures have been devised, which cannot be defeated, to detect the tap. The electromagnetic pulse after a nuclear explosion will cause massive destruction to electronic equipment in the vicinity, but nonconductive optical cable will not allow high currents to flow, thus minimizing damage.

Let us explore some of the negative aspects of fiber optics. One of the largest disadvantages is that fiber optics does not lend itself to bus systems because coupler losses are approximately that of 1 km of fiber. Some progress has been made to reduce these losses. Switches or fiber-optic relay technology is in its infancy, and losses across optical contacts are equivalent to approximately 2 km of fiber-optic cable.

Fiber optics is not well suited to use in analog transmission. For example, LEDs often require linearizing circuitry. This was also true for lasers but great strides have been made to alleviate these problems.

Nuclear radiation in the vicinity of fiber-optic cable will blacken the glass. New radiation-resistant fiber has been developed but the cost is still prohibitive for industrial uses.

Cost is one of the prime concerns of industrial users of fiber optics. CATV has kept broadband copper cable system costs low. However, several countries are experimenting with fiber-optic CATV cable plants. The Japanese are installing pilot link for two-way communication that will accommodate not only CATV but telephone and other additional services. One of the major obstacles to this service in the United States and other countries is political rather than technological.

Cyclic Redundancy Checking

Cyclic redundancy checking (CRC) is a technique for detecting errors. The method is presented here because integrated-circuit manufacturers have implemented it in monolithic circuits. Depending on the polynomial generator, CRC circuits will detect burst errors 18 bits in length for the common 16-bit integrated-circuit version and burst errors over 18 bits with an accuracy of 99 percent.

Two methods of CRC checking of message traffic are presented. The first technique is a rather simplified approach showing the arithmetic for performing the task. The second method is more representative of how the hardware actually performs the task.

CRC Method 1

$G(X) = X^5 + X^4 + X + 1,$ message to be sent

$G(X) = 110011$ (binary form)

$G(X) = 33_{HEX}$ (hexadecimal form)

$P(X) = X^4 + X^3 + 1,$ polynomial generated for CRC checking

$P(X) = 11001_B = 19_H$

$$\frac{X^4 G(X)}{P(X)} = Q(X) + FCS$$

This process shifts the message $G(S)$ 4 bits to the left prior to division and allows a 4-bit FCS.

$$\begin{array}{r} 21_H \\ 19_H \overline{)330_H} \longleftarrow G(X) \text{ shifted 4 bits} \\ \underline{32_H} \\ 10_H \\ \underline{19_H} \\ -9_H \end{array}$$

$$F(X) = X^4 G(X) + \text{FCS}$$
$$\text{FCS} = |\text{remainder}| = 9_H$$
$$F(X) = 339_H$$

The incoming message, $F(X)$, is checked for errors.

$$\frac{F(X)}{P(X)} = \frac{G(X) + \text{FCS}}{P(X)}$$

```
          21_H
   19_H ) 339_H
          32_H
          19_H
           19_H
           19_H
           00  ←——— Zero remainder indicates no errors
```

CRC Method 2

The transmission integrity of a received message is determined by the use of a frame check sequence (FCS). The FCS is generated by a transmitter, inspected by the receiver, and positioned within a frame in accordance with Figure 5–17. The procedure for using the FCS assumes the following:

1. The k bits of data that are being checked by the FCS can be represented by a polynomial $G(X)$. For example,
   (a) $G(X) = 10100100 = X^7 + X^5 + X^2 = X^2(X^5 + X^3 + 1)$
   (b) $G(X) = 00\ldots010100100 = X^7 + X^5 + X^2 = X^2(X^5 + X^3 + 1)$
   (c) $G(X) = 101001 = X^5 + X^3 + 1$
   In general, leading zeros do not change $G(X)$ and trailing zeros add a factor of X^n, where n is the number of trailing zeros.
2. The address, control, and information field (if it exists in the message) are represented by the polynomial $G(X)$.

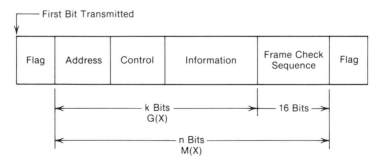

Figure 5–17

3. For the purpose of generating the FCS, the first bit following the opening flag is the coefficient of the highest-degree term of G(X), regardless of the actual representation of the address, control, and information fields.
4. There exists a generator polynomial P(X) of degree 16 having the form $P(X) = X^{16} + X^{12} + X^5 + 1$.

Generation and Use of the FCS

The FCS is defined as a one's complement of the remainder, $R(X)$, obtained from the modulo 2 division of $F(X)$ by the generator polynomial $P(X)$.

$$\frac{X^{16} G(X) + X^k(X^{15} + X^{14} + X^{13} + X^{12} + X^{11} + X^{10} + X^9 + X^8 + X^7 + X^6 + X^5 + X^4 + X^3 + X^2 + 1)}{P(X)}$$

$$\frac{X^{16}G(X) + X^k(X^{15} + X^{14} + \ldots + X + 1)}{P(X)} = Q(X) + \frac{R(X)\ FCS}{P(X)}$$

The multiplication of $G(X)$ by X^{16} corresponds to shifting the message $G(X)$ 16 places and thus providing the space of 16 bits for the FCS.

The addition of $X^k(X^{15} + X^{14} + \ldots + X + 1)$ to $X^{16}G(X)$ is equivalent to inverting the first 16 bits of $G(X)$. It can also be accomplished in a shift register implementation by presetting the register to all ones initially. This term is present to detect erroneous addition or deletion of zero bits at the leading end of $M(X)$ due to erroneous flag shifts.

The complementing of $R(X)$ by the transmitter at the completion of the division ensures that the transmitted sequence $M(X)$ has a property that permits the receiver to detect addition or deletion of trailing zeros that may appear as a result of errors.

At the transmitter, the FCS is added to the $X^{16}G(X)$ and results in the total message $M(X)$ of length $K + 16$, where $M(X) = X^{16}G(X) + FCS$.

The receiver can employ one of several detection processes, two of which are discussed here. In the first process, the incoming $M(X)$, assuming no errors [i.e., $M^*(X) = M(X)$], is multiplied by X^{16}, added to $X^{k+16}(X^{15} + X^{14} + \ldots + X + 1)$ and divided by $P(X)$.

$$Qr(X) + Rr(X)/P(X) = \frac{X^{16}[X^{16}G(X) + FCS] + X^{k+16}(X^{15} + X^{14} + \ldots X^1 + 1)}{P(X)}$$

Since the transmission is error free, the remainder, $Rr(X)$, will be 0001110100001111 (X^{15} through X^0).

$Rr(X)$ is the remainder of the division

$$\frac{X^{16}L(X)}{P(X)}$$

where $L(X) = X^{15} + X^{14} + \ldots + X + 1$. This can be shown by establishing that all other terms of the numerator of the receiver division are divisible by $P(X)$. This will be done later.

Note that $\overline{FCS} = R(X) = L(X) + R(X)$. Adding $L(X)$ to a polynomial of the same length is equivalent to a bit-by-bit inversion of the polynomial.

The receiver division numerator can be rearranged to

$$X^{16}[X^{16}G(X) + X^k L(X) + R(X)] + X^{16}L(X)$$

It can be seen by inspecting the transmitter generation equation that the first term is divisible by $P(X)$; thus the $X^{16}L(X)$ term is the only contributor to $Rr(X)$.

The second process differs from the first in that another term, $X^{16}L(X)$, is added to the numerator of the generation equation. This causes a remainder of zero to be generated if $M^*(X)$ is received error free.

Implementation

A shift register FCS implementation is described in detail here. It utilizes "ones presetting" at both the sender and the receiver, and the receiver does not invert the FCS. The receiver thus checks for the nonzero residual $Rr(X)$ to indicate an error-free transmission. Figure 5–18(a) is an illustration of the implementation. It shows a configuration of storage elements and gates. The addition of $X^k(X^{15} + X^{14} + \ldots + X + 1)$ to the $X^{16}G(X)$ can be accomplished by presetting all storage elements to a binary value of 1.

The one's complement of $R(X)$ is obtained by the logical bit-by-bit inversion of the transmitter's $R(X)$.

Figure 5–18(a) shows the implementation of the FCS generation for transmission. The same hardware can also be used for verification of data integrity upon data reception.

Before transmitting data, the storage elements, $X_0 \ldots X_{15}$, are initialized to 1. The accumulation of the remainder, $R(X)$, is begun by enabling A and thereby enabling gates G2 and G3. The data to be transmitted go out to the receiver via G2, and at the same time the remainder is being calculated with the use of feedback path via G3. Upon completion of transmitting the k bits of data, A is disabled and the stored $R(X)$ is transmitted via G1 and I_1 while G2 and G3 are disabled. The I_1 provides the necessary inversion of $R(X)$.

At the receiver, before data reception, the storage elements, $X_0 \ldots X_{15}$, are initialized to 1's. The incoming message is then continuously divided by $P(X)$ via G3 (A enabled). If the message contains no errors, the storage elements will contain 0001110100001111 ($X_{15} \ldots X_0$) at the end of the $M^*(X)$.

Figure 5–17(b) is an example of the receiver and transmitter states during a transmission of a 19-bit $G(X)$ and a 16-bit FCS.

The implementation of the FCS generation and the division by $P(X)$ as described are used as an example only. Other implementations are possible and may be utilized. This technique only requires that the FCS be generated in accordance

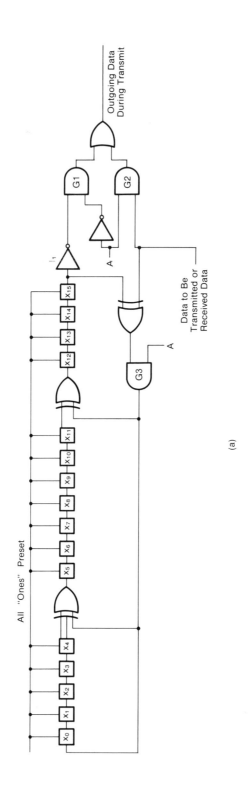

(a)

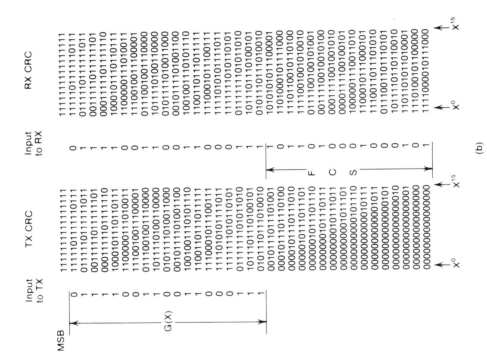

Figure 5–18 (a) A logic implementation of a CRC circuit. (b) An example table of FCS generation by the CRC circuit.

with the Rules 3.5 and 12.1 of the American Bureau of Standards and that the checking process involve division by the polynomial $P(X)$. Furthermore, the order of transmission of $M(X)$ is the coefficient of the highest-degree term first and, thereafter, in decreasing order of powers of X, regardless of the actual representation of fields internal to $M(X)$.

Transmitter Encoding

Transmission may be encoded for security purposes, to enhance the quality of data at the receiver end, or to superimpose data and clock information on a single waveform. The encoding discussed in the following paragraphs will address the latter (i.e., to superimpose clock and data on a single baseband waveform).

Manchester coding is a technique that has been used on disk drives previously. Figure 5–1 gives a diagram of a Manchester-coded waveform. The method of encoding is shown in Figure 5–19(a). Encoding is shown with a simple logic

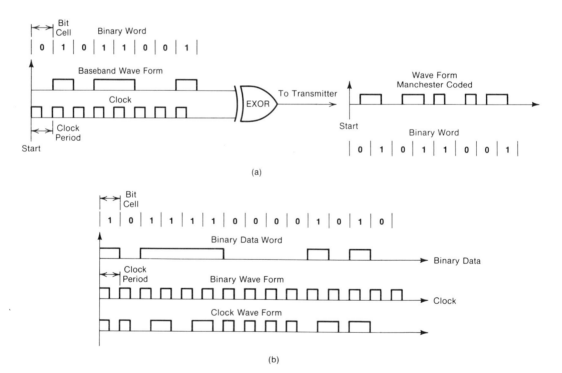

Figure 5–19 (a) Manchester encoder and waveforms. (b) Differential Manchester encoder and waveforms.

exclusive-OR gate (EXOR); it is implemented using a single IC package. The operation is as follows: Each bit is a cell, as shown in Figure 5–18(a). When the clock period is compared with the bit cell and the inputs are not identical, the EXOR produces a low output. For the opposite situation, (i.e., the two cell inputs are identical), the EXOR output is high. Therefore, note that when a binary zero occurs the EXOR produces a high during the last half of the bit cell, and for the binary 1 case, the clock and bit cell are identical over the first half of the cell.

Differential Manchester decoding is shown in Figure 5–19(b). The logic is not shown because it is a more difficult scheme to encode. An encoder is shown in reference [5]. The encoder functions as follows: When a binary 1 bit cell is compared with the clock period, transition will occur in the center of the bit cell only. If a binary 0 bit cell is compared with the clock period, transitions occur at the beginning and center of the bit cell.

The two self-clocked waveforms presented have other attributes that allow methods of transmitter-to-receiver signaling. For example, a clock pulse can be suppressed during transmission of these self-clocked waveforms. This is called a code violation. As a result, a period larger than 1 bit time will occur. As will be discussed in the next chapter, the receiver can determine if the beginning or end of a packet is being received. The decoding circuitry will be a little more complex because it must replace the missing clock pulse.

Another attribute of these self-clocked waveforms is that they prevent dc wander in receivers. The waveform has no dc component. For binary waveforms with large numbers of zeros or ones, the dc component can saturate the receiver, causing large numbers of errors in reception.

Several integrated-circuit manufacturers make Manchester encoder and decoder circuits in a single package. For transmission rates of 10 Mbits/s or less, fiber-optic transmitters and receivers are available with these features. All that is required is a clock and data input to the transmitter, and at the receiver, clock, clock data, and data are all sometime present at the output pins.

Receiver Decoders

The receivers extract clock and data from the incoming waveform as shown in Figure 5–20. The logic block is dependent on the encoding scheme used.

Note that in Manchester-coded waveforms the phase of the waveform is

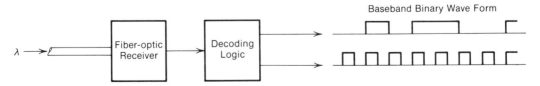

Figure 5–20 Manchester decoder block diagram.

necessary for decoding. If the medium is extremely noisy, the phase information may not be extracted easily; therefore, large errors will result. Differential Manchester is sensitive to transition that may be detected in the presence of noise. But errors in one bit can cause an error during the next bit interval, which doubles the error rate.

When SNRs are large enough to detect phase differences in the waveform, Manchester transmitters are a good choice; but if SNR is low, differential Manchester is superior. Many other codes exist that have attributes for various applications, but these two codes have been heavily implemented by industry.

Multilevel encoding techniques that use multiple levels to increase data rates are discussed in reference [7]. Signal analysis is also presented in that text.

The hardware available to design the physical layer of a LAN is quite diverse. In this text only the highlights have been examined. Prior to a physical-layer design, an assessment should be made to determine whether the LAN should be baseband or broadband based on data diversity. After this decision is made, the second question is, which is more suitable, analog or baseband? As one may observe, each decision is made to further narrow the scope until a technological approach is determined. To save time as well as money, several iterations of the system design should be made until the physical-layer design specification can be written with confidence.

Multiplexers

This discussion will address various types of time-division multiplexers, with a particular emphasis on statistical types. One technique that has been the most common for many years is the fixed-time-slot type. The nodes of a network have dedicated time slots, and data from each node must only be entered into the slot assigned to its node. This scheme is very inefficient; for bursty data from a computer only the time slots assigned can be used. The majority of the other nodes may not be active but the computer must wait its time slot each time before data transmission can occur. Channel utilization may be only 10 or 15 percent of total capacity.

For the time-division statistical multiplexer (TDSM), time slots are based on demand. If, for example, data from a computer are to be transmitted and the other nodes are inactive, the computer may burst all the data. But for the other situation, where many nodes are transmitting data, the computer and nodes will each get access to particular time slots; this will prevent the computer from monopolizing the channel resources. When only a few nodes are active (request transmission facilities), the channel data rate is high. As more nodes become active, frame size decreases so that individual node transmission rates appear to decrease. But if a limit on the number of active nodes is designed into the TDSM, then a minimum transmission rate will be ensured.

A comparison of common transmission techniques is shown in Figure 5–21.

Physical Layer

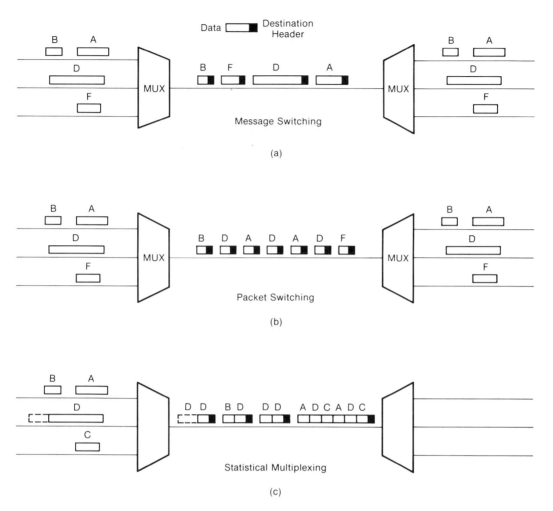

Figure 5–21 A comparison of message switching, packet switching, and statistical multiplexers.

Note, in Figure 5–21(a), that the messages are transmitted in their entirety, and if a short node message is transmitted behind a file transfer (a long string of data), the node data will have a long delay. The packet-switched data [Figure 5–21(b)] allows long messages to be broken into small packets of data. Long and short messages are interleaved, which prevents large transmission delays as compared to the subsequent techniques. The TDSM, however [Figure 5–20(c)], adjusts the frame size to the traffic situation, which implies that this device must have intelligence (micro-processor or computer).

Data-flow control in TDSMs is necessary to prevent errors in transmission

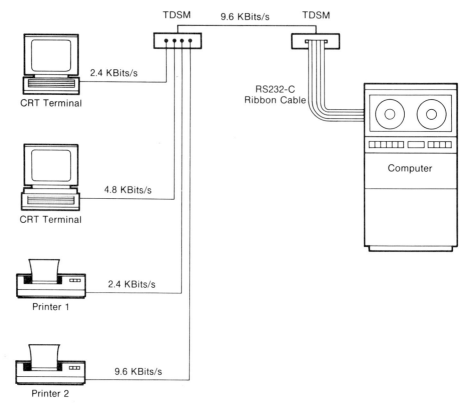

Figure 5–22 Typical situation requiring flow control equipment with a time-division statistical multiplexer.

or lost data. Figure 5–22 is a typical situation requiring data-flow control. If all devices connected to the transmission medium were transmitting simultaneously, the total accumulative transmission rate would be twice the line capacity, even if the TDSM will overflow, and data will be lost. Therefore, the TDSM must have some method to stop the data flow.

One method of stopping data flow to the TDSM to is provide an RS232-C connection between the TDSM and the implements it is connected to. When an implement is going to send data to the TDSM, it must send a request to send RTS on pin 4 of the RS232-C interface. The TDSM must respond with a clear to send CTC, pin 5; this handshaking is necessary before transmission can occur. When the TDSM buffer begins to fill to near capacity, the CTC line can signal the device to stop transmission.

If the RS-232 connections are not implemented in the TDSM design, which

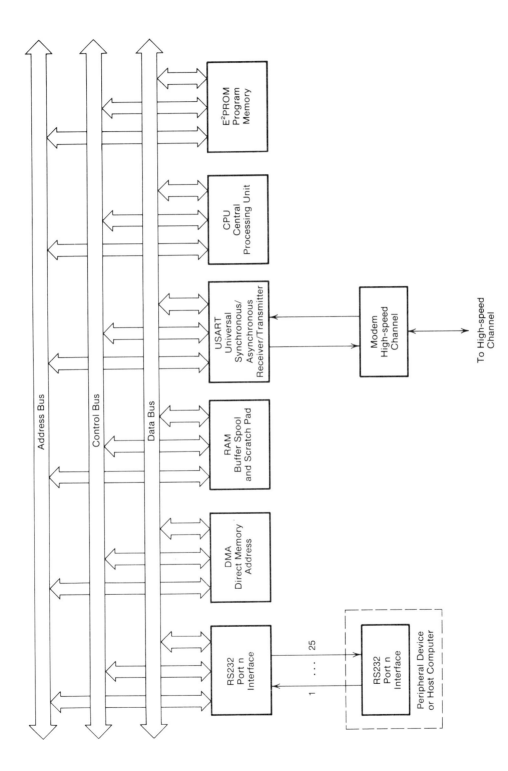

Figure 5–23 Block diagram of a statistical multiplexer.

is the case for high-transmission-speed networks, then in-band signaling would be necessary. This technique requires the TDSM to send a ASCTT X-OFF word to halt data transmission. Band signaling can use any type of code words that the channel controller will acknowledge. Most TDSMs have a dedicated microprocessor, and programs may be installed to accommodate a variety of channel configurations.

Another control technique is output control; that is, the output of the network is monitored (this can be a device or special monitor), and when traffic flow is approaching the limit of capacity, a halt of TDSM transmission is initiated.

Another type of multiplexer is the switching multiplexer. As was indicated previously, a statistical multiplexer assigns link bandwidth upon demand. The switching multiplexer assigns computer ports on demand and bandwidth on demand. It performs two levels of multiplexing.

The TDSM and switching multiplexers will be considered further in Chapter 7. The reason for this decision is that they involve multiple terminal and host computer connections. The multiplexers perform network control function due to the on-board intelligence.

Figure 5-23 is a block diagram of a TDSM. Data are received via the RS-232 interface, which is indicated as port n in the figure, implying several may be present. As data enter the port, they are converted from serial to parallel and stored in the RAM via the DMA channel. The central-processing unit will arrange the data in the transmission format. When the USART buffer is ready for data, a block move will be set up by the CPU and the transfer will be completed. The USART may have a DMA controller on board for the larger systems and it may consist of an entire circuit board, or, for small-scale systems, the USART may be a single IC and the CPU will handle data transfers. Also shown in Figure 5-23 is a single-port RAM. But in larger systems the RAM may consist of many circuit cards that are dual ported (i.e., it may receive data while data are being accessed by the CPU). This technique requires a more elaborate bus scheme, not shown here. An E^2PROM (electrically erasable programmable read only memory) is shown as the program memory, which may be altered when reconfiguration of the TDSM is necessary.

The block diagram shows only the functionality but not the true complexity of any particular TDSM. The equipment may range from a simple, single-board design solution to a rack-mounted minicomputer-based unit.

From this introduction, the reader should have gained some insight into configuring the cable plant hardware. The next ISO layer will delve deeper into the controllers that manage the physical layer and communicate with the upper layers.

PROBLEMS

5-1. Using Fourier analysis, draw the spectral plot for the waveform shown in Figure P5-1.

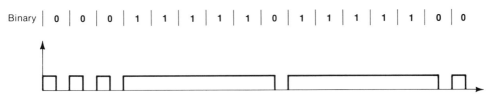

Figure P5-1

5-2. Derive a general equation using Fourier Analysis for binary waveforms. Bit times are designated as ΔT.

5-3. Show the proof that for OOK modulation the spectral plot of a binary signal is shifted up in frequency. See Figure 5–2(a) and (b). The proof should be general.

5-4. Draw a spectral plot of Equation 5–10.

5-5. Do a fiber-optic loss budget calculation for a link with the following specifications:

Maximum terminal distance	4 km
Cable attenuation	1.6 dB/km
Connector loss	0.7 dB
Switch insertion loss	1.5 dB
Number of switches between transmitter and receiver	2
Splice loss	0.1 dB/splice
Maximum number of splices	2
Code format NRZ	
BER worst case	10^{-9}
Gain margins	≥ 5 dB
Assume transmitter is operated at half-power	

(a) Which transmitter is adequate, LED or laser?
(b) What is the sensitivity requirement for the receiver? (This will be dependent on the transmitter selected.)

5-6. Calculate the rise time budget for the link in Problem 5–5 given the following data:

Transmitter rise time	50 ns
Transmission rate	1 Mbaud
Detector rise time	20 ns
Wavelength of operation	820 nm
Spectral width	40 nm
Numerical aperature, NA	0.20

5-7. For the link parameters given in Problems 5–5 and 5–6: (a) Is the system bandwidth or attenuation limited? (b) How can the limit be extended for the limiting condition?

5–8. Calculate the gain for the preamplifier A_1 and main amplifier A_2 for a fiber-optic video distribution system. The SNR must be 46 dB for an adequate display. A PIN diode is used in the detector with $\gamma = 0.5$ µA/µW. The modulation index of the incoming signal is 40 percent. Neglect I_d. $B_n = 6$ MHz and $R_L = 1$ kΩ.

REFERENCES

[1] Misha Schwartz, *Information Transmission Modulation and Noise,* McGraw-Hill Book Co., New York, 1980, pp. 274–277.

[2] D. Fink, *Electronics Engineers Handbook,* 11th edition, 1982, McGraw-Hill Book Co., New York, pp. 21–59 to 21–71.

[3] L. J. Giacoletto, *Electronics Designers Handbook,* McGraw-Hill Book Co., New York, 1977, Section 6.

[4] Digital Equipment Corporation, Maynard, Mass.; Intel Corporation, Santa Clara, Calif.; Xerox, "The Ethernet," A Local Area Network Data Link Layer and Physical Layer Specifications, Digital Equipment Corporation, Maynard, Mass.

[5] D. G. Baker, *Fiber Optics Design and Applications,* Reston Publishing Co., Reston, Va., 1985.

[6] Technical Staff of CSELT, *Optical Fibre Communications,* McGraw-Hill Book Co., New York, pp. 789–799.

[7] John C. Bellamy, *Digital Telephone,* John Wiley & Sons, Inc., New York, 1982, pp. 169–188.

6
DATA-LINK LAYER

The data-link layer provides data flow control across the physical layer. It decodes address information within the data stream and provides address encoding. This layer often provides error detection and corrections. The data-link layer provides the rules needed to allow the physical layer to communicate efficiently.

There are roughly three types of protocols: no cooperation, limited cooperation, and total cooperation. Examples of each are the following: no cooperation can be best represented by the ALOHA scheme; this technique allows a node to transmit without regard to the other nodes in the system. The collision when two or more nodes transmit is detected after the transmitting nodes fail to receive an acknowledgment of the message (after a delay period) from the receiver or if the transmitting node does not receive the transmitted data at its own receiver. Repeated collisions are avoided from the transmitting nodes because the delay before retransmission is controlled by randomly selected delays. These delays are due to random-number generators designed into the controllers. Slotted ALOHA is a form that has packets of uniform length. Once a node has seized the channel, it may transmit free of collision for the duration of the slot. This latter ALOHA is an example of a limited-cooperation protocol. A collision may occur during the channel seizure.

Total-cooperation protocols are characterized by no-collision transmission. Two examples are the token-passing ring and slotted line. The token-passing ring allows only the node that passes the token to transmit. After transmission, the sending node relinquishes the token and the next node desiring to transmit must capture it. When no nodes wish to transmit, the token is passed around the ring. The other technique, known as the slotted line, dedicates a time slot for each node. As the time slot passes the node, the packet is inserted with the destination address and the receiving node can determine the source of the data by the time

Data-Link Layer

slot. As can be observed, total cooperation between nodes is required to prevent any collisions.

The performance of noncooperative protocols is the poorest, and the total-cooperative protocol has the best performance. The complexity of the circuits required also follows this same relationship (i.e., noncooperative protocol is less complex than fully cooperative types).

Elementary Protocol

Several different types of protocols will be examined in the following paragraphs and a comparison of their performance is given. Some are popular and have been implemented in integrated-circuit form. Each LAN has certain attributes that must be stated before an analysis can be made.

The following assumptions are made about the LANs:

1. Each network has N nodes.
2. The message traffic is of a bursty nature.
3. Message traffic is generated by all nodes with equal probability.
4. The message packets are of equal length.
5. The worst-case delay is considered (i.e., the delay between the nodes separated by the largest distance).
6. All transmission is broadcast throughout the network.

Previously, the packet was defined, but only superficially. The packet is a unit in which data-link layers communicate. It is an organized assemblage of bits. As previously mentioned, the physical layer converses with other nodes in bits. The packet is composed of the address information (i.e., source and destination), control bytes, data, the FCS word, and sequencing numbers. A flag or delimiter is inserted at the head end and end of each packet to form a frame. Figure 6–1 depicts several different types of packets.

The first is an Ethernet packet. Prior to transmission of the synchronization bit, the Ethernet controller sends a preamble, which is a 64-bit pattern of alternating ones and zeros, except for the last two bits of the sequence; they are both ones. The preamble allows all the receivers to synchronize before the data-link layer begins transmission. The transceivers sense activity on the medium with a carrier sense signal. A synchronization bit in the first bit position of the packet allows the first bit from the receiver decoder to be invalid. If a double zero is encountered during the preamble, the packet will not be forwarded to the data-link layer. The frame must be retransmitted by the sending node. Normal operation again resumes on the subsequent frame. Ethernet frames do not have flags. Ethernet uses a single bit for recovering the clock at the beginning of a frame and the absence of transitions to detect the end. Data are Manchester encoded; therefore, even strings of ones or zeros will cause transitions, and idle channel conditions are depicted by no transitions.

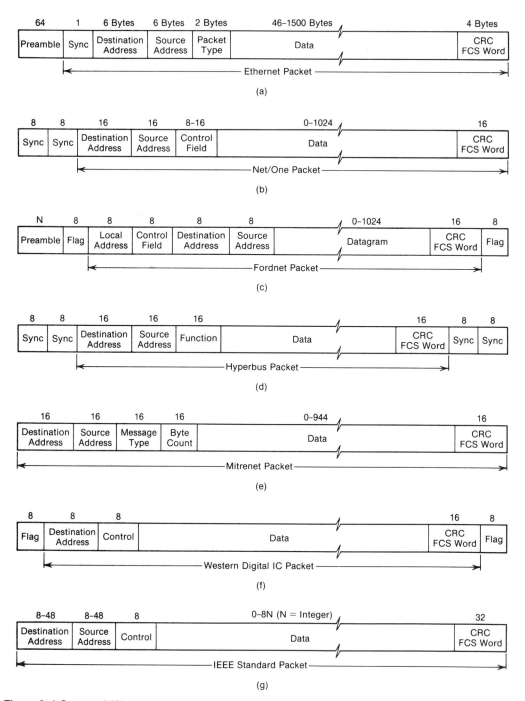

Figure 6-1 Common LAN packet and frame format.

NET/ONE has two synchronization bytes that function similar to delimiters at the beginning of a packet. The end of the packet is sensed as in Ethernet (i.e., the absences of transitions).

FORDNET has a preamble similar to Ethernet for receiver synchronization, with flags at the beginning and end of the frame to increase integrity, because in single-flag frames an error of one bit in the flag will cause the loss of the entire frame. When data errors can be tolerated, the CRC is disabled, which allows errors to occur; they may be corrected at higher ISO layers.

MITRENET uses a byte count in the frame header that determines the frame length. Errors in the frame byte count cause improper frame delimiting. The other fields are similar to those in previously discussed packets.

The proposed IEEE standard relieves the data-link layer protocol of the synchronization and framing, but a sublayer is included between the physical medium and the data-link layer to provide this service.

The last packet diagram is of particular interest because it is implemented in itegrated-circuit form. This particular one operates only in the token-passing mode. This protocol is compatible with the ISO (high-level data-link control) HDLC, the IBM (synchronous data-link control) SDLC, the ANSI (advanced data communication control procedure) ADCCP, and the CCITT X.25 protocols. All these protocols are bit-oriented rather than byte-oriented (i.e., the frames have an arbitrary number of bits rather than bytes). Later in the chapter, this controller will be discussed in detail. The discussion here is to acquaint the reader with some of the different packet and frame architectures.

Preambles are needed because some of the hardware is composed of circuitry that must synchronize before transmission to allow clock extraction circuits to stabilize prior to sending the packet. In other networks, the synchronization is combined with the flags (delimiters) to allow synchronization due to the flags. These preambles and flags are not part of the packet but are used by the physical layer to stabilize the network facilities.

Addressing schemes are quite varied. Let us examine some of the addressing techniques. Ethernet has two types of physical addresses. The first type is the multicast (broadcast) type where all bits in the destination address field are ones. The second type is specific destinations address, which is unique for each implement; the physical address of the equipment is displayed on it in hexadecimal form.

NET/ONE has a 16-bit address field that allows up to 65,536 addresses, as does MITRENET and Hyperbus. FORDNET and the Western Digital controller have only 8-bit addressing fields (256 addresses). But some of these address fields have provisions for extended addressing. For example, the Western Digital controller address can be extended by programming it. The controller, when programmed for extended addressing, will examine the least significant bit (LSB) of the address field. If it is zero, the next byte is an address byte. This process continues until an address byte with an LSB of one is detected, which terminates the address field (see Figure 6–2).

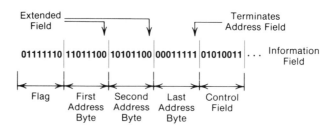

Figure 6–2 Western Digital WD1933 protocol header.

The IEEE has a similar scheme for extended addressing, except the address field has 2, 4, or 6 bytes. If the most significant bit (MSB) of the 16-bit address field is a one, then the next 2 bytes are address fields that allow 24^7-1 addresses or about 300 trillion. The minus one is because a broadcast address is required.

Source addresses are included in most of the packets shown in Figure 6–1, because if an acknowledge (ACK) or not-acknowledge (NACK) is required by the transmitting node the source address is needed. Another technique for checking the accuracy of the transmitted data is to allow the transmitting station to check incoming data at its receiver (for ring-type networks). For this situation, the source address should be that of the sending station or node; therefore, everything transmitted must be identical to what is received or an error in transmission has occurred.

Note that the Western Digital controller packet does not have a source address because this controller has a token-passing protocol. When the receiver of the transmitting station receives flags and an address, it is the only transmitter that could have sent the information; collisions are ever present, except possibly when the token is lost or damaged. With some modifications, the source field can be added as part of the address field. (This method was used in an architecture the author was involved with.)

The Ethernet-type field is uninterpreted at the data-link layer. It is reserved for use at higher levels to identify client-layer protocol associated with the frame. Ethernet has no provisions for layers above the third; therefore, an uninterpreted field will assist in supervision of the lower levels.

The Western Digital controller has a control field, but it must be acted on by a local microprocessor. This external microprocessor is programmed to perform the necessary tasks to permit the combination to emulate data-link layer functions. For example, the WD1933 can emulate the SNA and X.25 data-link-level protocols (the header is shown in figure 6–2). A microprocessor is needed to decode the control word that determines frame type, ACK/NACK, and other functional features of the protocol.

Let us examine the control word of the frame (see Figure 6–3). An information frame is designated [Figure 6–3(a)]; it has a zero in the first bit position.

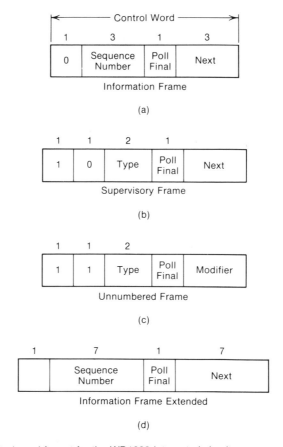

Figure 6–3 Control word format for the WD1933 integrated circuit.

The next part of the control word is the sequence number. Each frame is given a sequence number during transmission in the range from 0 to 7. The receiving station (remote station) sends the sequence number in the NEXT field if the frame is received correctly. This is commonly called piggybacking the acknowledge. All the protocols adhere to one of two conventions for multiple frame transmissions instead of piggybacking each ACK separately. The receiver sends the sequence number of the last correct frame or the first frame not received. Either technique is adequate, but it must be consistent.

The poll or final bit is used when a computer or concentrator is polling a group of terminals. The computer or concentrator sends out a poll message with poll or final bit = 1; the terminal or other peripheral sends out a frame with poll or final = 1. The computer or concentrator sends out a series of frames to the terminal (poll or final = 0) until the last frame is sent (poll or final = 1 or end of transmission).

TABLE 6–1
Control Word Supervisory and Unnumbered*

Function	Acronym	Control Field
Connection request	SABM	1111 p 100
Confirm response	UA	1100† 110
Reject connection response	DM	1111† 000
Frame ACK	RR	1000 p/f $n_1n_2n_3$
Reset request	FRMR	1110 f 001
Receive not ready	RWR	1010 p/f $n_1n_2n_3$
Disconnect request	DISC	1111 p 101
‡Selective reject	SR	1011† $n_1n_2n_3$

*See Figures 6 3(b) and 6–3(c).

Notes: p, poll bit

†, copied from corresponding command

f, final bit

p/f, poll or final bit (depends on application)

‡calls for retransmission of frame specified; HDLC and ADCCP only allow this frame type

The supervisory and unnumbered control words are shown in Table 6–1. The supervisory frames are defined below:

Supervisory Frames [Figure 6–3(b)]

Type 0 Frame ACK (RR): This frame is used for acknowledgment when no reverse traffic is available to piggyback an ACK from receiving station to transmitting station.

Type 1 Reject (REJ): This frame is used to indicate that an error in transmission has been detected. The $n_1n_2n_3$ field indicates the first frame in the sequence that was not received correctly. The transmitting station or node must retransmit that frame and other outstanding frames.

Type 2 Receive Not Ready (RNR): The receiver notifies the transmitting node that it acknowledges all frames up to but not including $n_1n_2n_3$. This type of control word is intended to halt the transmitter because it has temporary problems such as buffer overflow. When the condition is corrected, the Receiver Ready (Frame ACK) frame will reestablish the connection.

Type 3 Selective Reject (SR): This type of frame allows for retransmission of a single frame, which is only present in HDLC and ADCCP protocols.

The next control frames to consider are unnumbered frames.

Unnumbered Frames [Figure 6–3(c)]

Disconnect (DISC): This allows communicating nodes to inform each other when they are discontinuing the communication channel for maintenance purposes.

Set Normal Response Mode: This command is a normal type that announces the presence of a node on the network. It forces all sequence numbers back to zero and is suitable for host and terminal communication. This command is not used in most symmetric LANs (i.e., LANs where all nodes have equal access to the network).

Connection Request (SABM): Table 6–1 depicts the control word; the acronym stands for Set Asynchronous Balanced Mode. This control word resets both node sequence numbers. SABME is the extended version of the SABM. The SABME control word is shown in Figure 6–3(d). This control word is used in the HDLC and LAPB procotols.

RESET Request (FRMR): This control word indicates that the frame has the correct checksum, but there are semantic errors, the frame is shorter than 32 bits, it is an illegal control form, an ACK was received outside the window, and so on, when the control word is in the frame.

Confirmed Response (UA): A 4-byte data field indicates an error condition. When this type of control frame is present, it has a special (unnumbered ACK) from (VA). Only one control frame is transmitted; therefore, the UA frame when received has no ambiguity.

Often other control words are generated that are used for housekeeping and control functions between nodes. These are transparent to upper layers.

The next item in a packet to consider is data. Packets may be received out of order, and others may require retransmission if this occurs; these are the characteristics of virtual circuit service, which is representative of most modern LANs. The other technique of message transmission is the datagram. The data are sent as a message; no attempts are made to retransmit erroneous messages. An example of a datagram is mail delivery. The letter is received as a total message and letters may be delivered to the receiving end out of order (i.e., letters 1 and 2 may be sent on two consecutive days, but letter 2 may arrive before letter 1). This type of service is also characteristic of telegraph service, which is considered a datagram. In this text, only virtual-circuit-type service will be considered.

After the data field, the next item in the packet to consider is the checksum or FCS word. A discussion of the mechanics for CRC checking was given in Chapter 5. The FCS word can be appended or checked as in Chapter 5, or the incoming frame FCS can be compared directly with an FCS word generated at the receiver. The technique described in Chapter 5 uses the same checksum on each frame, while the latter technique will have a different checksum for each frame. The CRC checks the data stream for burst errors depending on the number of bits in the CRC polynomial. See Chapter 5 for more information on CRC techniques.

The last items in the frame are the flags that signify the end or beginning of the frame. In Ethernet, with no flag at the end of a frame, the frames sense inactivity on the channel for a timeout period (i.e., no transitions occur).

The previous discussion has centered on the mechanism of data-link-level communication between nodes. The issue of how the channels are initiated must be considered next and whether the protocol is noncooperative, limited cooperative, or fully cooperative.

The first type to be considered is the ALOHA type (noncooperative). The ALOHA protocol was originally used on radio networks and was later adapted for

LAN use. These LANs allow nodes to transmit at any time (no cooperation between nodes). The circuitry is simpler to implement but the performance is poor. When two nodes transmit at the same time, the receiver notes the errors but does not send an acknowledgment. The transmitting nodes do not receive acknowledgments; after a timeout period, they must retransmit the data. As one can imagine, during heavy traffic periods a great many collisions occur, thus causing large delays before a message can reach its destination correctly. These long delays will cause losses in throughput.

If the network access is such that the transmitting node can receive its own transmission, then collisions can be detected by the receiver when they first occur after the network delay. The transmitting nodes will stop transmitting if a collision occurs for a random period and begin again if no traffic is sensed on the channel. This type of protocol is commonly called carrier sensed multiple access (CSMA). There is more than one classification for CSMA protocols. Nonpersistent CSMA is the type just described. Persistent CSMA is characterized by nodes that begin trying to transmit as soon as no transmission is sensed. Note that this technique senses collisions at the receiver of the transmitting node, which requires long delays before the collisions can be sensed due to propagation.

Suppose the collision can be sensed at the transmitting node during transmission (i.e., the collisions are detected immediately). This is called appropriately CSMA/CD, the CD being collision detect. This is the technique used for implementing the Ethernet protocol.

A technique used by IBM, which is an IEEE 802 standard, is the token ring. If a node wishes to transmit, it must have the token. When no stations are transmitting, the token is passed from node to node. This technique requires total cooperation between nodes. In a variation on token passing, CSMA/CD is used when the channel is quiet (no transmission is sensed). When a station finally seizes the channel, it transmits the data frame and appends a token after the flag on the last frame sent. Any node desiring to seize the channel only need wait until the token appears and seize it. The station that has the token can then transmit.

The slotted ring requires each station on a ring-topology network to have a unique time slot. It may transmit only during its time slot. This type of network needs a master station to manage and assign time slots. Throughput will suffer with this technique, but the reliability of this type of system is very high; therefore, a trade-off can be made.

A discussion of network topology is reserved for Chapter 7, but some performance issues will be addressed in the following paragraphs with a comparison of protocols. For example, ring topology may be implemented with token passing, CSMA/CD, CSMA, slotted ring, or register insertion. Bus topology may be implemented with CSMA, CSMA/CD, token, or broadband bus with CSMA/CD. A performance comparison of some of these techniques will be presented here. For further information on the subject, see reference [1].

Highlights of the calculations used in reference [1] will be presented here. Equation 6–1 defines a parameter that has a large impact on the channel utilization. The equation expresses the ratio of the worst-case length of the data path to packet length.

$$a = \frac{R \times D_p}{L_p} \frac{\text{bit}}{\text{bit}} \qquad (6\text{-}1)$$

where R = data rate in bits/second
D_p = worst-case propagation delay
L_p = maximum packet length

An example of the calculation is shown in Example 6–1.

Example 6–1

Given a fiber-optic link with a 1-km worst-case distance between nodes, a packet length of 128 bytes, and a transmission rate of 10 Mbits/s, calculate the value a.

Fiber-optic propagation delay = τ_d

$$\tau_d = \frac{n}{c} \approx 5\frac{\text{ns}}{\text{m}}$$

$$D_p = \tau_d l = 5 \text{ }\mu\text{s at 1 km}$$

$$a = .049$$

Note: $a = D_p/\tau_p$, where τ_p = packet transmission time.

Another important feature is channel utilization, which is given by Equation 6–2. The value of a is a parameter used in the calculations. This relationship is

$$U = \text{channel utilization} = \frac{\text{throughput } (T_r)}{\text{data rate } R}$$

$$T_r = \frac{L_p}{D_p + (L_p/R)}$$

$$U = \frac{L_p}{D_p R + L_p}$$

$$= \frac{1}{(D_p R/L_p) + 1} \qquad (6\text{-}2)$$

$$= \frac{1}{a + 1}$$

As a decreases, the value of U approaches 1, which implies that utilization is 100%. This occurs when the packet length approaches infinity or worst-case delay approaches zero; this of course never occurs in practical systems.

The utilization analysis given is an intuitive approach. Several items are not considered because they tend to make the model much more complex and cumbersome to derive. Neglected in the analysis are time delays due to collisions, acknowledgment packets, waiting time for tokens, slotted-ring waiting time for slots, and register insertion delay. These items are characteristic of CSMA/CD, token bus and ring, slotted ring, and register insertion LANs. Before resorting to more complex calculation, the token-passing bus and ring will be examined and the CSMA/CD protocol.

Let us examine the calculations required to predict throughput performance of a token-passing network (i.e., either bus or ring):

$$\tau_c = \text{average time for one cycle}$$
$$\tau_p = \text{average transit time for a data packet}$$
$$\tau_T = \text{average token passing time}$$
$$\tau_r = \text{throughput}$$
$$N = \text{number of stations}$$
$$\tau_c = \frac{\tau_p}{\tau_c} = \frac{\tau_p}{\tau_p + \tau_T}$$

For networks with $a < 1$ as previously defined (i.e., $\tau_p/\tau_p < 1$)

$$\tau_c = \tau_p + \frac{\tau_p}{N}$$

Then

$$\tau_r = \frac{1}{1 + (a/N)} \quad (6\text{–}3)$$

If the networks have values of $a > 1$, then Equation 6–4 predicts throughput τ_r.

$$\tau_c = \tau_D + \frac{\tau_D}{N}$$

Then

$$\tau_r = \frac{\tau_p}{\tau_D + (\tau_D/N)} = \frac{1}{a(1 + (1/N))} \quad (6\text{–}4)$$

As would be expected, throughput improves when propagation delays are short and packets are long. Let us examine Equation 6–4 for a moment. As the number of stations on the link is increased, the throughput will become dependent on the ratio of τ_p/τ_D or $1/a$. The equation will continue to function in a correct manner (i.e., decreasing delay or increasing packet size will improve throughput).

Data-Link Layer

Next to be considered is the CSMA/CD throughput. These calculations are a bit more involved than for token passing and therefore a few definitions are in order.

τ_s = time slots on the medium

$\tau_s = 2\tau_d$

P_s = each station transmits during an available time slot with the probable P

The time within the medium consists of two types of intervals. These are the transmission interval, $a/2$, and the collision interval.

For a calculation of the contention the probability, P_A must be calculated (i.e., the probability that only one station acquires the medium). This can be depicted using the binomial probability function of the form shown in Equation 6–5. P_A is given by Equation 6–6.

$$f(x) = \binom{N}{1} p^x q^{N-x} \qquad (6\text{--}5)$$

$$P_A = NP(1 - P)^{N-1} \qquad (6\text{--}6)$$

Differentiating Equation 6–6 and setting the derivative equal to zero, the maximum value of P_A can be calculated.

$$\frac{dP_A}{dP} = N(1 - P)^{N-1} - NP(N - 1)(1 - P)^{N-2} = 0$$

$$P = \frac{1}{N}, \text{ solving the equation for } P \qquad (6\text{--}7)$$

Substituting the value in Equation 6–7 into Equation 6–6, the original expression for P_A, the result is

$$P_A = \left(1 - \frac{1}{N}\right)^{N-1}$$

This is the maximum probability of P_A.

Next the mean length of the contention interval must be estimated. The assumptions are that the time slots on the link have alternating intervals with no transmission or may have a collision at transmission intervals. Equation 6–8 describes this function.

$$P(w) = P_A(1 - P_A) + 2P_A(1 - P_A)^2 + \cdots + KP_A(1 - P_N)^k$$

$$= \frac{1 - P_A}{P_A}, \text{ the value of } P(w) \text{ at series convergence} \qquad (6\text{--}8)$$

$$\text{CSMA/CD } \tau_r = \frac{\tfrac{1}{2}a}{\tfrac{1}{2}a + P(w)} = \frac{1}{1 + 2a[(1 - P_A)/P_A]}$$

The equations for the token bus and ring are evaluated for $N \to \infty$ as shown for Equations 6–3 and 6–4.

$$\text{Token ring and bus } \tau_r \lim_{N \to \infty} = 1, \quad a < 1$$

$$\tau_r \lim_{N \to \infty} = \frac{1}{a}, \quad a > 1$$

For the CSMA equation, P_A must be evaluated for $P_A = [1 - (1/N)]^{N-1}$, the maximum value.

$$P_A = \frac{[1 - (1/N)]^N}{[1 - (1/N)]} = \frac{1}{2!} - \frac{1}{3!}\left(1 - \frac{2}{N}\right) + \frac{1}{4!}\left(1 - \frac{2}{N}\right)\left(1 - \frac{3}{N}\right) + \cdots$$

$$P_A \lim_{N \to \infty} = \frac{1}{2!} - \frac{1}{3!} + \frac{1}{4!} - \frac{1}{5!}$$

Then P_A can be written

$$P_A = 1 + (-1) + \frac{(-1)^2}{2!} + \frac{(-1)^3}{3!} + \frac{(-1)^4}{4!} + \cdots$$

or

$$P_A = \frac{1}{e}\Big|_{N \to \infty}$$

Then

$$\tau_r = \frac{1}{1 + 2a(e - 1)} = \frac{1}{1 + 3.436a} \tag{6-9}$$

For token-passing networks, throughput performance increases as the number of nodes increases, and for $a > 1$, the network performance becomes dependent on a. A good rule to follow is try to make the packet size large as compared to delay, which may not always be possible. This same rule applies to CSMA/CD networks. As more stations appear on the network, the likelihood of a collision increases and performance will suffer.

Various curves showing performance are presented in Figure 6–4(a) and (b). Figure 6–4(a) shows throughput as a function of a. Note from the curves that if $a = 0.1$, increasing the number of stations on a link improves throughput for token passing and decreases throughput for CSMA/CD links. This was stated previously for Equations 6–3 and 6–4, but here it is shown graphically. Figure 6–4(b) shows throughput as a function of the number of nodes on the network for token passing and CSMA/CD for two values of a.

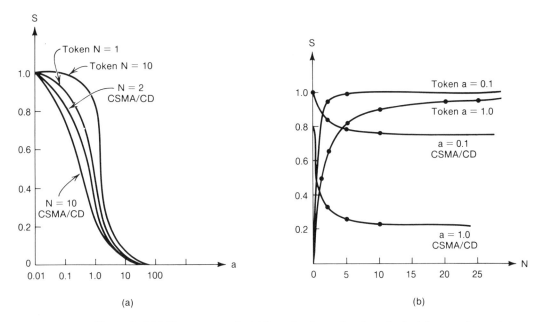

Figure 6–4 (a) These curves depict throughput as a function of a for token passing and CSMA/CD. (b) These curves depict throughput as a function of N for token passing and CSMA/CD.

A study by the IEEE 802 Local Network Standards Committee [2] produced results that are shown graphically in Figure 6–5. From Figure 6–5(a) and (b), it can be seen that the actual data rate is almost at the same level as the channel data rate (channel transmission rate) for both token-passing protocols. The reductions in transmission rate are fairly drastic for the CSMA/CD bus protocol because a great deal of time is lost servicing collision.

Lightly loaded networks, as shown in Figure 6–5(c), indicate token rings and CSMA bus have almost identical performance for larger packets, but the token bus has relatively poor actual data rates. For smaller packets, the data rate is reduced even further for the token bus.

These curves are useful when making a decision on which type of protocol is best for a particular application because they show how throughput, traffic, protocols, packet size, and data rate affect networks.

For a comparison of various ring topologies, see references [3], [4], and [5]. Excerpts from these references would have been presented here except that some of the analyses are tentative. To make a comparison of ring networks is a rather momumental task because of the large number of variables and is beyond the scope of this book. For example, using small values of a may yield high throughput for one protocol and the opposite for another. Maybe a change in this variable

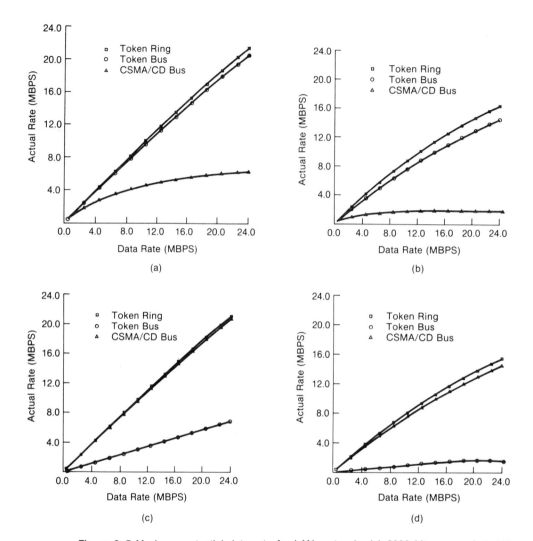

Figure 6–5 Maximum potential data rate for LAN protocols. (a) 2000 bits per packet, 100 stations active out of 100 stations total. (b) 500 bits per packet, 100 stations active out of 100 stations total. (c) 2000 bits per packet, one station active out of 100 stations. (d) 500 bits per packet, one station active out of 100 stations.

may have a large impact on other parameters of other protocols. To make a fair comparison of networks, the designer must fix some of the parameters (e.g., minimum acceptable actual transmission rate, number of stations, and packet length). After some of the parameters are fixed, certain protocols and topologies will drop out as undesirable based on performance.

Throughput and channel efficiency will be examined to determine the impact of the header, retransmission, number of ACK bits, and the like, on these performance criteria. One obvious question is, how many bytes should be in the data field for optimum efficiency? The calculations will provide the reader with the necessary equation to answer this question.

The first protocol to consider is a stop-and-wait type. No piggybacking is allowed (i.e., the control field will not have an ACK type word). The variables are as follows:

A = acknowledge frame bits
C = channel capacity in bits per second
D = data bits per frame
P_E = probability of a bit error
$F = D + H$
H = number of bits in the header and flags
I = interrupt and service time + propagation delay
P_A = probability of a lost or damaged ACK frame
P_0 = probability a data frame is lost or damaged
R = number of retransmission
ΔT = timeout interval
V = channel utilization or efficiency
W = window size
$I = \tau_i + \tau_d$

The first analysis is of transmission of a frame without any errors. Intuitively, let us consider the time required to send a frame, service an interrupt, and prepare to send an ACK frame. This time is calculated using Equation 6–10.

$$\tau_{TF} = \frac{F}{C} + I = \tau_F + \tau_i + \tau_d = \text{total time to service frame}$$

$$\tau_{TA} = \frac{A}{C} + I = \tau_A + \tau_i + \tau_d = \text{total ACK service time}$$

$$\tau_T = \tau_{TF} + \tau_{TA} = \tau_F + \tau_A + 2\tau_i + 2\tau_d = \text{total data frame and ACK service time} \quad (6\text{–}10)$$

$$\tau_D = \frac{D}{C} = \text{data transmission time}$$

$$V = \frac{\tau_D}{\tau_T} = \frac{\tau_D}{\tau_F + \tau_A + 2\tau_i + 2\tau_d}$$

Equation 6–10 can be expressed as a ratio of bits as shown in Equation 6–11. An alternate form is equation 6–12.

$$V = \frac{D/C}{(F/C) + (A/C) + 2I} = \frac{D}{F + A + 2IC} \quad (6\text{–}11)$$

$$V = \frac{D}{H + D + A + 2IC} \quad (6\text{–}12)$$

When transmission of a frame occurs and each time the frame has errors, then the total number of retransmission bits is $R(F + TC)$. The value $F + TC$ will continue until an ACK frame is sent or until the node stops trying to transmit. A successful transmission will require the number of bits shown, with utilization depicted in Equations 6–13 and 6–14.

$$B_T = \text{unsuccessful tries} + \text{one succesful try}$$
$$= R(F + TC) + (D + H + A + 2IC)$$
$$V = \frac{D}{R(F + TC) + (F + A + 2IC)} \qquad (6\text{–}13)$$
$$V = \frac{D}{R(D + H + TC) + (D + H + A + 2IC)} \qquad (6\text{–}14)$$

The first form does not show the complete breakdown of the frames and is more compact. The latter form will allow the reader to differentiate the equation and find relationships between data capacity and other variables.

But before the equation can be completed, the R value must be further examined. The mean number of retransmissions per frame must be estimated based on the probability of successful transmission of data and acknowledgment. The development of these equations is shown next. Note that to complete a transmission of data the ACK must also be considered.

$$P_s = (1 - P_D)(1 - P_A) = \text{probability of successful transmission}$$
$$P_f = (1 - P_s) = \text{probability of failure}$$
$$n = \text{number of transmission attempts needed}$$

Then $P^{n-1}(1 - P_f)$, with $n - 1 = $ number of retransmissions, expresses the number of transmissions needed to transmit the frame. The number of transmissions per frame is

$$T_R = \frac{1}{P_s}$$

The number of retransmissions, therefore, is dependent on the probability of failure:

$$R = T_R P_f = \frac{P_f}{P_s}$$

Substituting the result of this equation into Equation 6–14 yields the following expression:

$$V = \frac{D}{(P_f/P_s)(D + H + TC) + (D + H + A + 2IC)} \qquad (6\text{–}15)$$

Data-Link Layer

For the condition $T \approx (A/C) + 2I$, the timeout, Equation 6–15 will reduce to Equation 6–16.

$$V = \frac{D}{(D + H + A + 2IC)[(P_f/P_s) + 1]}$$

$$= \frac{D}{D + H} \times \frac{P_s}{P_f + P_s} \times \frac{1}{1 + [(A + 2IC)/(D + H)]}$$

Then

$$V = \frac{D}{D + H} \times P_s \times \frac{1}{1 + [CT/(D + H)]} \quad (6\text{–}16)$$

For an error occurring in the frame for both ACK and data frames, now consider the probability of an error in each bit P_E. Then an error in the ACK and data frame is represented by $(1 - P_E)^{H+D}$ and $(1 - P_E)^A$, respectively. Then $P_s = (1 - P_E)^{H+D} \times (1 - P_E)^A$ is substituted into Equation 6–16, and let $A = H$ (i.e., the ACK frame is the length of the header).

$$V = \left(\frac{D}{D + H}\right)(1 - P_E)^{H+D}(1 - P_E)^H \left(\frac{D + H}{D + H + CT}\right) \quad (6\text{–}17)$$

Taking the derivative of U with respect to D and setting it equal to zero will result in an optimum value for D or D_{opt} if the roots are found. The optimum D_{opt} is derived as follows:

$$\frac{\delta V}{\delta D} = \left(\frac{1}{D + H + CT}\right)(1 - P_E)^{2H+D} - \left[\frac{D}{(D + H + CT)^2}\right](1 - P_E)^{2H+D}$$

$$+ \left(\frac{D}{D + H + CT}\right)\frac{\delta(1 - P_E)^{2H+D}}{\delta D}$$

$$0 = (D + H + CT)(1 - P_E)^D - D(1 - P_E)^D$$

$$+ D(D + H + CT)\frac{\delta(1 - P_E)^D}{\delta D}$$

$$0 \approx (H + CT)(1 - P_E)^D + [D^2 + D(H + CT)](1 - P_E)^D \ln(1 - P_E)$$

Then

$$0 = [D^2 + D(H + CT)]\ln(1 - P_E) + (H + CT)$$

$$D_{\text{opt}} = -\frac{H + CT}{2} \pm \sqrt{\frac{(H + CT)^2}{4} - \frac{(H + CT)}{\ln(1 - P_e)}} \quad (6\text{–}18)$$

$$= \frac{H + CT}{2}\left[\sqrt{1 - \frac{4}{(H + CT)\ln(1 - P_E)}} - 1\right]$$

The expression in Equation 6–18 is an equation for optimizing D; note that the negative radical has been ignored when finding the roots of D_{opt} because the D can only have positive values. If the probability of a bit error is small, which is the usual case (common values are $P_E = 10^{-8}$ or 10^{-9}), $\ln(1 - P_E) \approx -P_E$ in the equation reduces it to the form of Equation 6–19.

$$D_{opt} = \frac{H + CT}{2}\left[\sqrt{1 + \frac{4}{P_E(H + CT)}} - 1\right] \qquad (6\text{--}19)$$

The term $\frac{4}{[P_E(H + CT)]} \gg 1$; therefore, the value for D_{opt} can be approximated by Equation 6–20.

$$D_{opt} \approx \sqrt{\frac{H + CT}{P_E}} \qquad (6\text{--}20)$$

An examination of Equation 6–20 reveals some important observations. As the error rate gets smaller, $P_E \to 0$, the optimum data field becomes infinitely large. As headers and timeouts become large, the frame size must increase to improve U, which is realistic.

High-level Data-link Control (HDLC) and Synchronous Data-link Control (SDLC)

SDLC is a subset of HDLC, which is defined in the ISO standards [6]. The difference between the two protocols is in the control field as was previously discussed in the first section of this chapter.

The transmitter and receiver must be synchronized using a clock derived from the data waveform (self-clocked) or a quasi-synchronous clock (i.e., a locally generated clock synchronized by the incoming data). The latter case will disallow the transmission of certain characters or will require bit stuffing. The two methods are represented in Table 6–2.

TABLE 6–2
Relationship between SDLC and Data Communication Equipment (DCE)

DCE	SDLC
Asynchronous modems (e.g., FSK and baseband): DTE clock	Useful, but a terminal with a synchronous data-link control (DLC) can often justify a higher-speed modem
Synchronous DCEs: DCE clock derived from data waveform	Feature of the design
Synchronous DCEs: DCE clock independent of data	Feature of the design

The first entry in Table 6–2 is similar to the technique used in the WD1933 IC, which emulates the SDLC/HDLC protocol (i.e., constant zeros on the line are represented by a plus signal for one-half the bit cell and zero for the second half of the cell). If the transmitter is sending zeros, a signal similar to the clock appears at the output. When binary ones are transmitted, the output stays high. Transitions are induced in the transmitter output for large strings of ones by inserting a zero after every five ones (bit stuffing). Bit stuffing can also be accomplished for the reverse condition (i.e., ones have the transitions and zeros remain at zero volts, then ones are needed to force transitions).

Errors in HDLC/SDLC protocols are detected by the CRC controller logic or by standard ICs such as the Signetics 8X01. With the CRC, undetected errors will be 1 frame in 20 billion. Errors only occur with a burst of 18 bits or more. Single errors are detected and retransmission is requested with a NACK by the receiver.

Error recovery in half-duplex operation is as follows: The frames are numbered at the transmitter so that the receiver can determine if a previously received frame was correct. Next, if all the frames are received correctly or incorrectly, the receiving node is required to send an acknowledge or not-acknowledge (ACK/NACK).

When an error is detected by the CRC, the receiver responds by sending the old frame number. If the last frame accepted is n and the frame number of the frame being checked is $n + 1$, the receiver responds by sending $n + 1$ ACK. If the frame number of the frame being checked is n, then a duplicate frame is sent and discarded. For a received frame with the number $n + 2$ and the last frame accepted is n, the receiver responds with a NACK.

The transmitter checks the response of the receiver for errors ACK/NACK. If an erroneous response is received, the most recent frame(s) sent is (are) retransmitted (i.e., frame n). If the response received is error free, it is either n or $n - 1$, with n being the last number transmitted. If n response is received at the transmitter, then $n + 1$ is transmitted, and when $n - 1$ is received at the transmitter, n is retransmitted.

Another situation to be addressed is lost frames due to short-term channel outages. Often during electrical storms, short-term brownouts may cause power supply dropouts or relay dropouts, which may cause entire frame loss. If frames are not received within some prescribed time, a request for retransmission is sent. The prescribed time is set by a software timer located in the data-link controller (DLC).

For full-duplex operation, each transmitting node sends independent sequence numbers. The error recovery is similar to two half-duplex operations. That is, functionally, node X is transmitting to Y and node Y is also transmitting to node X.

When a node is recovering from errors, the SDLC protocols provide a number of retries and a pause. The number of prescribed retries and the pause time are selected by the user. If the number of prescribed retries and pauses are exhausted, the transmitting node conveys a message to the operator(s) of the node that the receiving node is inoperative. This is passed upward through the layer hierarchy, which allows the operator(s) to take appropriate action.

The link-level protocol must account for the following:

1. The channel bandwidth, which is usually not the same as physical system bandwidth
2. Channel propagation delay
3. Message generation patterns
4. Channel bit error probability
5. Maximum and minimum frame size
6. Length of ACK frame
7. Maximum number of sequence numbers
8. Maximum number of ACKs to be sent per message
9. Maximum ACK-delay timeout interval
10. Window size
11. Node buffer availability

ARPANET

ARPANET protocol sets up a buffer prior to transmission of the message. A complete frame is sent. It takes approximately 20 ms and an interrupt is then generated. The transmission hardware sends three characters, synchronize (SYN), data-link escape (DLE), and start of text (TeXt STX.) These three characters define the beginning of a frame. The transmitter empties the frame buffer and automatically sends DLE and end of text (ETX) characters, followed by a 24-bit CRC checksum (FCS word). After this sequence of transmission, the hardware sends SYN characters.

The ARPANET hardware performs character stuffing. When a DLE, ETX sequence occurs in the text, the transmission hardware inserts an extra DLE. The code embedded in the text now reads DLE, DLE, ETX. The receiver removes the extra DLE at the receiving node. A frame length field in the header and the normal CRC checksum prevent any larger error rate due to character stuffing.

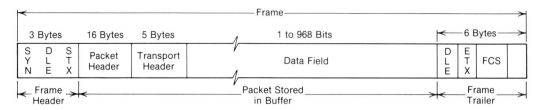

Figure 6–6 ARPANET frame format.

The ARPANET frame is described pictorially in Figure 6–6. For star-type networks, the packet waits in queue until the routing algorithm decides on which output line to use. The packet is then assembled into a frame by the hardware and transmitted on the selected line.

The software within the node multiplexes eight logical full-duplex channels onto every physical line; satellite links have 16 channels due to their inherent long delays. Each of these channels uses a stop-and-wait protocol (i.e., the sender waits for an ACK for each frame before proceeding with transmission). Each data-link controller maintains a table with 3 bits per logical channel to determine if an outstanding packet is on the channel, the sequence number of the next packet expected, and the sequence number of the next packet to send. When a channel is selected, it is marked busy and transmission begins. After completing transmission and receiving an ACK, the channel condition changes from busy to idle.

Some important properties of ARPANET are the following:

1. *Full-duplex paths:* The two nodes communicating have identical buffering and transmission and receive facilities.
2. *Sequence of data transfer:* Data being transferred must remain in sequence when received; therefore, the receiver must have a reassembly capability and the transmitter must have disassembly processing.
3. *Transparency of data:* Data must have any sequence of ones and zeros allowed (i.e., even group of bits that simulate flags); this implies, as previously mentioned, character or bit stuffing.
4. *Signaling:* The nodes must be capable of responding to inquiries both in band and out of band. In band is via the normal data stream, and out of band does not use normal data characters but some form of mechanism for defeating bit or character stuffing.
5. *Flow control:* The network is capable of reducing the flow of data in the network to prevent congestion.
6. *Error detection:* The network is equipped with CRC.
7. *Flexibility of switching:* Switching is accomplished by modifying the packet address field, and addresses of nodes can be easily changed.
8. *Interface independence:* The operation of nodes is independent of the physical, optical (fiber optics), or RF link, or broadband cable, that is, the medium in general.

Terminals

ARPANET terminals are of two varieties. The first is the nonintelligent type; it operates at from 75 to 600 bits/s; this terminal lacks the capability of responding to protocol features of the packets. The second type of terminal is intelligent; it can respond to all protocol features, and it also operates at high transmission rates of 1200 to 19.2K bits/s. See reference [7] for more information on ARPANET terminals.

SNA and X.25

A large part of these protocols was covered when frame structure was discussed in Figure 6–1. No further discussion will be presented here except to refer the reader to reference [8] prior to designing any X.25 circuits.

DECNET

The Digital Equipment Corporation network is unique compared to some of the other protocols. In this section, the subtleties will be explored. The protocol is called digital data communications message protocol (DDCMP).

DDCMP provides the user with a sliding window, with a maximum capacity of 255 packets outstanding. The maximum packet size is 16,383 bytes, as compared to the other protocols, such as ARPANET and Ethernet, which have maximum packet sizes of 1008 and 1500 bytes, respectively.

DECNET has the capability of handling half-duplex, full-duplex, and multidrop lines. In the multidrop line situation, one node is designated the master and all other nodes are slaves. All communications must go through the master station.

A pictorial of a DECNET packet is shown in Figure 6–7. The start of header (SOH) has the bit pattern 1000 0000, which is similar to a flag. The count field will give the length of the data field in bytes, a minimum of 1 and a maximum of 16,383.

The synchronization bit is used to synchronize both sides of the line if necessary. The select bit is used during half-duplex or multidrop operation. The bit is set when no further transmission from the sender is necessary; then the receiving node can begin transmission. On multidrop lines, this bit, when set, informs the slave nodes that the master is ready to receive.

The ACK is a field reserved for piggybacking an acknowledgment. The number of the last packet correctly received appears in this field. A sequence field provides a sequence number for the packet (0 to 255). The address field is used to direct the packet to the correct node.

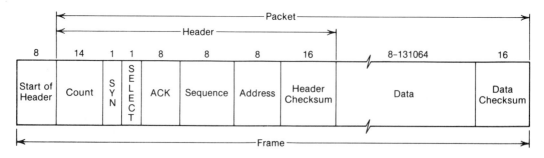

Figure 6–7 DECNET frame format.

A checksum is calculated using CRC-16, $P(X) = X^{16} + X^{15} + X^2 + 1$, which checks the header for errors. This CRC is necessary because an inadvertent error in the count field would possibly allow several packets to appear as data before the error is detected. Then, instead of a single packet in error, multiple packets would be received in error. The other protocols have timeouts or end-of-packet delimiters to prevent this situation from occurring.

The data field is self-explanatory, with the exception that no character or bit stuffing is required; this simplifies the hardware. The data checksum is only for the data part of the frame, while the header checksum only examines the SOH up to the end of the address field.

DECNET packets have no control field; therefore, some means must be provided for control. DECNET provides control packets for this purpose. The control frames are ACK, NACK, REP, START, and STARTACK. The ACK control packet is only necessary when reverse traffic is not available to piggyback the ACK. NACKs are used to indicate a packet in error or that all packets up to the one in error were sent correctly. When packets are transmitted, the receiving node must respond with either an ACK or NACK.

The REP packet sent after a long packet requires the receiver to send an ACK or NACK and prevents the transmitter from automatically sending the packet again before the status of the packet is known. The last two control packets, START and STARTACK, are used to initialize the transmit–receive pair.

Ethernet

The frame and packet description have been discussed previously, but some of the control functions need to be addressed here. When an Ethernet transceiver is not transmitting, it will monitor the physical channel traffic. If the channel is busy, the carrier sense signal is acted on by the station controller and defers any transmission until the last bit of the frame passes. Interframe spacing has a minimum value of 9.6 μs; the value may be increased at the expense of lower throughput.

Collision handling and detection were only superficially described previously; they will be further classified here. Contention for the channel occurs until a station has seized the channel. Slot time determines the dynamics of the collision-handling process. This parameter describes the following important aspects of collision handling:

1. It is an upper bound on network acquisition time.
2. The upper bound frame fragments, due to collisions, are determined by the slot time.
3. This parameter is the scheduling quantum for retransmission.

To fulfill these three functional requirements, the slot time is larger than the round-trip physical layer propagation time plus the data-link layer maximum jam time. The calculation is as follows:

Round trip delay = 45 μs = 450 bits
Maximum jam time = 4.8 μs = 48 bits
Total slot = 49.8 μs = 498 bits

The slot time is defined as 51.2 μs or 512 bits/64 bytes. The jam time (more than one station attempting to transmit) occurs at the beginning of transmission, and the condition is maintained for a minimum of 32 bit times but no more than 48 bit times.

When a collision occurs, the station attempting to transmit will try 16 times (i.e., the initial plus 15 retries). Each retry is attempted after a random delay period generated by a "truncated binary exponential backoff," which is a random generator. The delay is an integral multiple of the slot time ($0 \leq I \leq 2^{10}$) range of the integer multiplier.

When frames are received, decapsulation is required. The process involves removing the packet, address recognition, FCS (frame check sequence) validation, and passing all the fields to the next layer in the hierarchy. This layer also filters out any damaged frames that occur due to collision or general errors in transmission.

Frames may have two types of errors (i.e., excluding bit errors). A frame may be longer than 1518 bytes or the length may not be an integral number of bytes. The data-link layer truncates any frames to the nearest byte, and errors of such frames are reported as alignment errors to the upper layers in the hierarchy.

Modeling the Ethernet data-link layer can be accomplished using a procedural model. A good presentation can be found in reference [9]. Due to lack of space, it is not presented here, but an overview is given.

Procedural Model Overview

The procedural model is written in Pascal; it is intended as the primary specification of the Ethernet data-link layer functions. The program is a modeling tool, and it must not be considered as a program to be executed in a computer to provide data-link layer control. The model functionality should be implemented with the appropriate hardware, firmware, and software combinations. The implementation must match the specification, the end result, and not the internal structure of the model. Pipelined models of processing frames in both transmit and receive modes are more realistic than the model shown in reference [9]. Again, this is not reflected in the internal structure of the model and should be considered.

The algorithms for the model execute due to concurrent processes: these algorithms collectively represent the Ethernet data-link control processes. Timing dependencies due to concurrent activity are resolved using the following methodology.

Data-Link Layer

External Events versus Processes

The algorithms are executed at least 50 to 100 times faster than external events; that is, the process never falls behind in its work or fails to process an external event. For example, when a frame is received, the data-link procedure receive frame is called well before the frame to be processed has arrived.

Processes versus Processes

Each process is structured to external events, and interactions between processes are dependent indirectly on external events. Therefore, two interacting processes must be independent of their execution time.

The concurrency of the model may reflect a certain amount of parallelism intrinsic to the implementation of this layer, but in actuality the amount of parallelism varies.

Protocol Specifications and Verification

The specifications provide a formal, umambiguous, functional description of what the protocol algorithm does; included is all essential descriptive material, and all that is not pertinent is omitted. Verification is a guarantee that the algorithm, such as a program version, is in consonance with its specification. Verification is similar to acceptance testing of a unit; to be positive, it meets some given specification. Even with formal specification, a possibility of error exists because some feature may have been overlooked.

The protocol is an algorithm that involves both the source and receiving nodes; this is why the execution of the algorithm appears to occur in parallel. Some of the classes of algorithms that we may attempt to specify and verify are parallel algorithms, involving parallel programs or processes, which offer the most difficulty. See reference [10]. Consequently, verification and specification of protocols are rather difficult. However, much progress has been made in this area. For example, layering of protocols has simplified these tasks. Specification and verification of each layer separately and keeping the interfaces at the layer boundaries well defined will reduce the complexity of these two tasks.

Formal techniques for specifying protocols include state diagrams, formal grammars, Petri nets, high-level languages and I-systems. See references [11] through [15].

A state diagram of a state machine is shown in Figure 6–8. This example will give the reader some insight into how a graph is developed from the machine functional description. Table 6–3 is a description of the state diagram function with a description of each function of the nodes.

TABLE 6–3
State Diagram Functional Description

Initial state, the sender transmitting the 0 frame; the receiver expects a 0 frame; transmission of the ACK frame is present in the channel; and the channel is correctly delivering the data to the receiver, transition 1, or the contents are lost, which is transition 0.

The transmitter is transmitting the 0 frame, the receiver expects the 0 frame, but the data are lost and eventually the sender times out, transition 7, and begins at the initial state.

In this state the sender is trying to send frame 0, the receiver is waiting for frame 1, and the ACK is accepted by the receiver, which means that after the transmitter finishes receiving the ACK(A) the sender will expect to transmit the 1 frame. If the contents of the channel are lost, transition 0 occurs.

If these next states occur, emanating from the previous one due to the channel losing its contents, then the channel has lost its contents and the receiver is expecting frame 1 because the ACK sent to the transmitter was not received. If a timeout occurs (transition 7), then frame 0 will again be sent by the transmitting station, the receiver expects frame 1, and frame 0 is in the channel. If frame 0 is sent correctly, a return to the previous state will occur, 01A, and for the situation when the channel loses its contents, a return to 01n will occur.

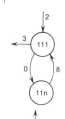

If in this state transition 2 has occurred, that is, the ACK was received at the sending station and frame 1 is being sent, frame 1 is expected at the receive station, and frame 1 is in the channel. If the frame is received correctly, transition 3 occurs, but if the frame is lost, transition 0, then the receiver will not send a ACK and the channel state will indicate a 1 frame is being sent again.

In this state, the sender is trying to transmit frame 1, the receiver expects frame 0, and the ACK is impressed on the channel. If the channel loses its contents, the machine will go into the 10n state until sender timeout occurs (transition 8). The sender will enter state 101 (i.e., transmitting station is sending frame 1, the receiver expects frame 0, and the ACK is in the channel). Transition 4 occurs if the ACK is accepted by the sender and frame 0 is transmitted. Transition 4 will return the communication back to the initial state.

Transitions 1, 2, 3, and 4 occur in sequence if the communication between transmitter and receiver is correct. They are summarized in Table 6–4.

One must keep in mind, when multiple frame sequences occur, that the protocol must determine what to do with frames out of sequence even if they are not allowed.

To properly analyze a protocol, one must ensure that no deadlocks occur. In logic circuitry, this same phenomenon occurs when a counter, due to an error, jumps into a nonallowed counting loop from which it cannot escape. A deadlock occurs when the protocol jumps to a disallowed state in which it cannot escape. When using a state graph, account for all states to ensure that no deadlock can occur.

Data-Link Layer

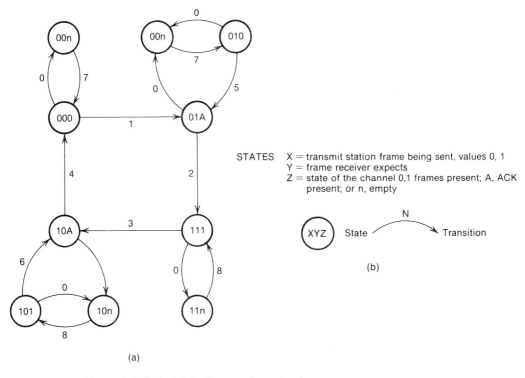

Figure 6-8 Typical state diagram of a protocol.

TABLE 6-4

	Receiver	Transmitter
Transition 1	Frame 0 accepted ACK being sent	
Transition 2		ACK accepted Frame 1 being sent
Transition 3	Frame 1 accepted ACK being sent	
Transition 4		ACK accepted Frame 0 being sent
Transition k*	Frame n accepted ACK_n being sent	
Transition $k + 1$		ACK_n accepted Frame $n + 1$ being sent

*The pattern can easily be extended to include a general case.

REFERENCES

[1] W. Stallings, "Local Network Performance," *TEEE Communications,* vol. 22, no. 2, Feb. 1984, 27–35.

[2] B. Stuck, "Calculating the Maximum Mean Data Rate in Local Area Networks," *Computer,* May 1983, pp. 72–76.

[3] M. T. Liu, W. Hilal, and B. H. Groomes, "Performance Evaluation of Channel Access Protocols for Local Computer Networks," *Proceedings COMPCON FALL 82,* 1982, pp. 417–426.

[4] W. Bux, "Local-Area Subnetworks: A Performance Comparison," *IEEE Trans. Commun.,* Oct. 1981, pp. 1465–1473.

[5] W. Bux and M. Schlatter, "An Approximate Method for the Performance Analysis of Buffer Insertion Rings," *IEEE Trans. Commun.,* Jan. 1983, pp. 50–55.

[6] "IBM Synchronous Data Link Control General Information," IBM Systems Library, order number GA27-3073, *An Introduction to SDLC.*

[7] Leonard Kleinrock and H. Opderbeck, "Throughput in the ARPANET Protocols and Measurement" *IEEE Trans. Commun.,* vol. COMM-25, no. 1, Jan. 1977, pp. 95–104.

[8] "Revised CCITT Recommendation X.25-1980," NCS Technical Information Bulletin 80-5, August 1980. Available from National Communications System, NCS-TS, Washington, D.C. 20305.

[9] Digital Equipment Corporation, Intel Corporation, Xerox Corporation, "The Ethernet: A Local Area Network: Data Link Layer and Physical Layer Specification," Version 2.0, Sept. 30, 1982.

[10] H. K. Berg and others, *Formal Methods of Program Verification and Specification,* Prentice-Hall, Inc., Englewood Cliffs, N.J., 1981.

[11] P. Zafiropulo and others, "Towards Analyzing and Synthesizing Protocols," *IEEE Trans. Commun.,* COMM-28, April 1980, pp. 651–660.

[12] J. Harangoso, "An Approach to Describing a Link Level Protocol with a Formal Language," *Proceedings of the Fifth Data Communications Symposium,* Salt Lake City, Utah, 1977, pp. 4–37 to 4–49.

[13] F. J. W. Symons, "Modeling and Analysis of Communications Protocols Using Numerical Petri Nets," Department of Electrical Engineering, University of Essex, England, Technical Report 152, May 1978.

[14] C. A. Ellis, "Consistency and Correctness of Duplicate Database Systems," *Proceedings of the Sixth ACM Symposium on Operating Systems Principles,* Purdue University, Lafayette, Ind., Nov. 1977; *ACM Operations Systems Review II,* 1977, pp. 67–84.

[15] A. Danthine and J. Bremer, "Modeling and Verification of End-to-End Transport Protocols," *Proceedings of the Computer Network Protocols Symposium,* Liege, Belgium, Feb. 13–15, 1978.

7
NETWORK LAYER

Virtual Circuits

One of the most important aspects of a local-area network with packet switching is that of the virtual circuit. If a single channel within a particular medium is shared by multiple users, the time-shared channel between any two users is a virtual circuit. These virtual circuits must have certain properties to be able to service the network layer. The following paragraphs provide a discussion of each of these attributes.

Data Transfer in Sequence

The data bits delivered to the network for transmission must maintain their order through the network. This implies that the network facilities must disassemble messages before transmission and reassemble them at the receive end.

Network Transparency

The network must maintain transparency to its users, and no groups of ones and zeros can be excluded from the data field of a packet. Data must be delivered error free between any two users. This condition implies checking for inadvertent flags generated within the data field.

Path Full-duplex Capability

The path must appear to be operating full-duplex between any two users (i.e., transmission must appear to be simultaneous). Thus the nodes in the network must

Network Layer

have identical network capability. For example, initiation of a connection and message buffering in one direction must have the same attribute for data in the other direction.

Control Signaling

The signaling must flow freely between users to maintain flow control. Status information must be conveyed between nodes to respond to user inquiries. Signaling can be produced (in band) as part of the user data stream or (out of band) outside the normal data stream.

Data-flow Control

The network must prevent congestion within the network during high-traffic-density periods. Traffic-routing algorithms are one example of flow control.

Error Control

All network transmission must have a negligible amount of errors. The network layer may demand retransmission of data with errors or perform error correction.

Interface Independence

The network layer must be independent of physical properties (electrical or fiber optic) of the node interfaces. It must also be consistent with logical data structures.

Address Flexibility

Network virtual circuit operation allows information to be exchanged between various user pairs by modifying user address fields within a message.

Datagrams

The next network service to consider is the datagram. When the facilities are shared and entire messages are transmitted, this service can be characterized by stock quotation, reservations, and point-of-sale terminal-type systems. Datagrams are characterized by the following set of characteristics.

Self-contained Messages

The information within the datagram is self-contained; that is, it does not depend on previous messages nor does it contribute information to any other message except the present one.

Datagram Identification

The datagram contains the destination address routing information and control; it is a complete entity.

Errors in Transmission

The destination node has no previous knowledge that the datagram is being sent. Any lost datagrams are not recovered or retransmitted due to a request from the destination node. Often duplicate datagrams may be sent by the transmitting node to ensure that the data have been conveyed to the destination node.

Sequencing

Datagrams are not sequenced since they are complete entities. Messages transmitted to a destination node need not arrive in order because they are not sequenced.

Uncontrolled

The originating node for datagrams will be advised of the datagram progress through the network, but it is uncontrollable after leaving the originating node.

A summary of the differences between datagrams and virtual circuits is given in Table 7–1.

If one examines the table, it can be deduced that the datagram is the more primitive of the two network services. One traffic control (routing algorithm) is the software responsible for deciding which output is to be used for transmission of an incoming packet.

TABLE 7–1
Comparison of Virtual Circuits and Datagrams

	Virtual Circuit	Datagram
Destination address	Only during initial start-up	Needed in every packet
Source address	Only needed at start-up	Not always needed
Error detection	Transparent to upper network layers	Done by the upper layers
Flow control	Provided by the network layer	Not provided by the network
Packet sequencing	Messages passed in order	No order required
Initial network setup	Required	Not possible

Many of these local-area networks may be required to run for years without problems; therefore, they must not only be robust, but the routing algorithm must be simple and precise. The topology must be flexible (i.e., addition of new implements on the network should not require changes in the software). As an example, if a LAN has been installed at an insurance company with networks using smart terminals as nodes, the addition of new terminals should not present a software problem to the previously installed terminals. Also, all nodes within the network should be treated equally (i.e., each can access the network with equal probability of success).

Routing Techniques

Two forms of control are used on routing algorithms. The first is a nonadaptive type where traffic is routed according to a fixed table; the second type is adaptive, and the algorithm will measure traffic flow and route traffic accordingly. This technique is also called *dynamic routing.*

Another method of routing, called *flooding,* uses a broadcastlike technique. The outgoing packets are sent to all lines except the line they were received on. Fiber-optic star couplers lend themselves well to this type of transmission. Star couplers are made rather simply. Several fibers are twisted together and fused. An input signal is reproduced at every port except the incoming signal port. A discussion of this component is deferred until later when star network connection is discussed.

One can imagine what flooding does to traffic flow when several nodes are transmitting simultaneously. Flooding has a limited amount of utility as compared to other routing techniques. Some networks need this attribute to function properly (e.g., a network distributing news, stock market results, tactical military networks (where redundancy is a necessity), and other similar networks when data must be distributed to all nodes). It is also possible to partially flood a network (i.e., distribute over only a part of the output lines). If this technique is used to forward data in the direction of the destination node and all other nodes are capable of forwarding the data toward the destination, the shortest route to the destination will be selected. However, these techniques all increase traffic density.

Directory routing is another technique of packet control and is the most common. The routing directory is set up by the operator with prior criteria used by the operator, such as shortest path, least delay, or light traffic volume.

Examples of routing tables are shown in Figure 7–1(a) and (b). Node 5 of Figure 7–1(a) will be used as an example. To convey data to destination node 1, nodes 2 or 6 must be used as intermediate nodes to pass the data. Either of these nodes in turn must pass the data on to node 1 or, if redundancy is required, both nodes 2 and 6 must pass the data to destination node 1.

The second routing table [Figure 7–1(b)] is determined by the shortest path. This type of network was examined in more detail in the analysis given in Chapter 4.

If minimum delay were the objective of the routing table and a node were assumed to have delay, then entries in the table must be examined for path length and the number of nodes data must pass through. The table in Figure 7–1(b) has

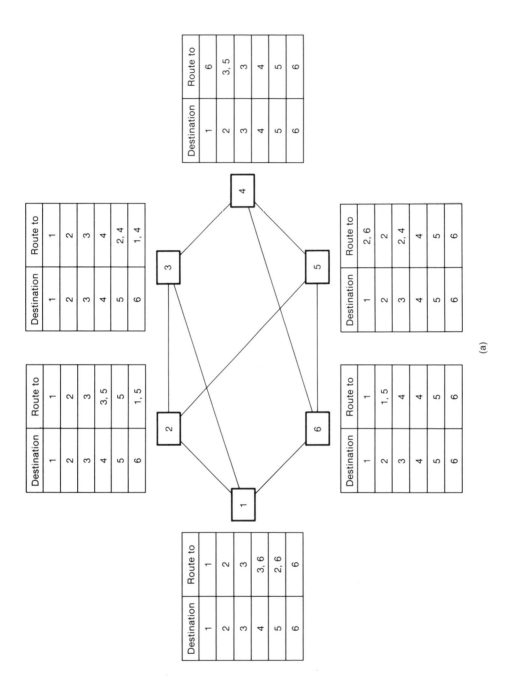

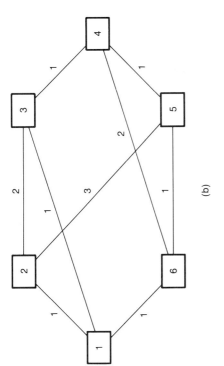

Figure 7–1 (a) Routing tables based on routes to reach a destination with no regard to physical network. (b) Routes are labeled with distances between nodes given in the routing table in part (c). (c) Routing table based on the shortest path from origin to destination node.

two entries for destination node 2 and originating node 5 labeled (6, 1), 2. This indicates the signal may travel from node 5 to 2 or from 5 to 6, 6 to 1, and 1 to 2. Each of these paths has equal length; however, if delay through the nodes is considered, only the path 5 to 2 can be considered as an entry. If several possible routes emerge as candidates to carry the message traffic, then a random-number generator can be implemented to select routing.

The discussion has dealt primarily with nonadaptive routing. The next logical step is to examine adaptive routing, which of course is much more useful than the nonadaptive case.

Adaptive routing has one immediate disadvantage over nonadaptive and that is the complexity of executing and updating the routing matrix in real time. However, the new generation of microprocessors will certainly overcome the computing power issue. These microprocessors have the computing power of minicomputers, and they may, in certain cases, be added in parallel to provide parallel processing. Large numbers of controllers will eventually be available that emulate specific network protocols, and these devices will no doubt have adaptive routing capability.

Now let us examine what is involved in adaptive routing. Some form of monitor is required to update network status, for example, traffic density over various routes, delay along all routes, and lastly the condition of all nodes along the route. One must also note that all these quantities are highly time dependent. Traffic density may increase and decrease very rapidly because of the bursty nature of the data. Delay that is purely due to propagation is very predictable, but when a node that processes data lies within the path of transmission, the service time of the node can vary depending on loading. Nodes within the network may become damaged or removed from service; these contingencies must be reflected in the network status.

Techniques for developing a routing table will be addressed next. The objective of any local-area network is to deliver packets of data in a rapid and efficient manner. When the network is lightly loaded, then minimum delay is the logical choice (shortest path), and data must pass through the fewest number of nodes (unless nodes act as repeaters). This last statement will be clarified further when ring networks are examined.

If the routing table is made adaptive and based on minimum delay, the node transmitting data must select the line to its nearest neighbor. This line condition must be known (i.e., the number of bits waiting in the queue, the transmission rate over the line, error rate, and the traffic density). When this information about its neighbor is known, the node can calculate the estimated delay to it. An estimate of nodal delay is made to all nodes other than the transmitting node's nearest neighbor, because the nearest node may have long queuing delays, lower transmission rates, higher bit error rates, or heavier traffic density than some of the other more distant nodes.

The transmitting node must exchange delay information with its neighbor. Thus the nodes within this network will have a microprocessor or specialized microprocessor-based controller to provide the necessary computing power to accomplish these various tasks and perform delay calculations.

The network in Figure 7–2 shows delay at each node. The delays in the table entries are all in milliseconds. Note that all the propagation delays are insignificant if queuing delays are present; therefore, the nodes will make decisions based on queues for a heavily loaded network. For a lightly loaded network where the queues are empty, the routing decisions will be made based on the propagation delay.

The routing tables shown in Figure 7–2 are incomplete because they do not show the delay to all the nodes from the originating node. Our analysis will start with node 5, and the other nodes can be filled in by the reader (see Problem 7–1).

The origination node 5 (Figure 7–3) has the minimum delays as shown. Let us examine each table entry. The route 5–1 gives the originating node as 5 and the destination node as 1. For the route shown from 5 to 1, the data must be processed by node 2, which delays the data by 20 ms. The route (node 5–4) can be explained in a similar manner; data will pass through node 2 when node 5 originates the data and node 4 is the destination with node 2 an intermediate node data must pass through. The third entry was already made in the table as a delay of 312.1 ms, but a shorter delay was found via node 3. Also, note that many other entries are possible, but they do not have the minimum delay. As an example, see Table 7–2.

TABLE 7–2
Table of Routes from Node 5 to 1

Route 5–1 = route 5–2 + route 2–1
route 5–2 + route 2–4 + route 4–1
route 5–3 + route 3–1
route 5–6 + route 6–4 + route 4–2 + route 2–1
route 5–3 + route 3–6 + route 6–4 + route 4–1
route 5–2 + route 2–4 + route 4–6 + route 6–4 + route 4–1

Table 7–2 can be further extended, but certain entries will become delay prohibitive due to nodal processing or queuing delay and, perhaps in some cases, propagation delay. As previously investigated in Chapter 4, the numbers of paths will soon become prohibitive and to determine these paths a large amount of computing power will be needed.

A tree can be used to generate all possible paths, as shown in Figure 7–4, which is a graphic representation of the paths from 5 to 1. The graph was generated by starting from node 5 and working always toward node 1. For example, trace the path to node 3; from node 3 a path progresses to node 1 and 6. But we are not interested in the path from 6 to 5; therefore, only the path to 4 is considered. After the paths to node 4 are traced, two paths to node 1 are apparent: routes

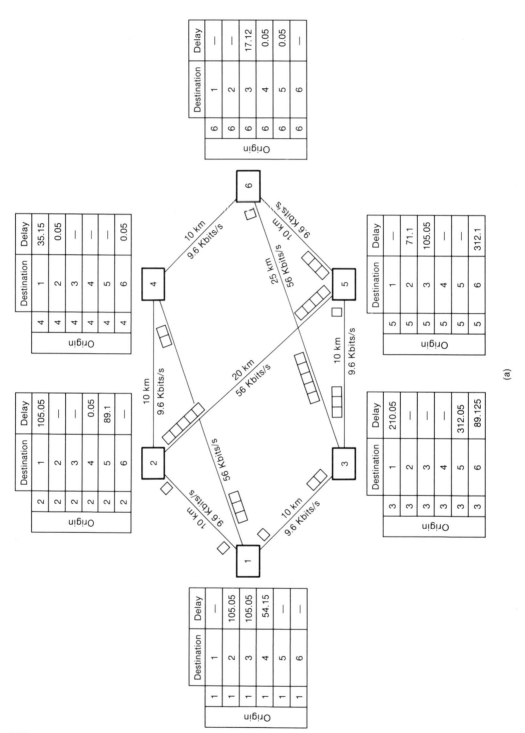

(a)

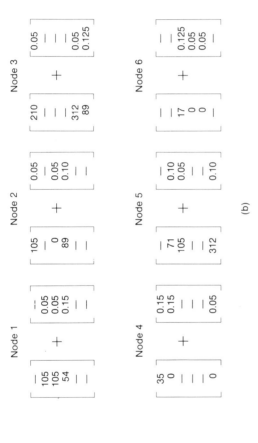

Figure 7-2 (a) Network with delays for each route shown. (b) Matrix representation for each node with the queue and propagation delays separated into two matrices.

```
                Node 5
                         Total
Origin   Destination    Delay
  5          1          146.30    ←——— Route 5-1 = route 5-2-4-1
  5          2           71.10
  5          3          105.05
  5          4           91.15    ←——— Route 5-4 = route 5-2-4
  5          5            —
  5          6          214.175   ←——— Route 5-6 = route 5-3-6

                                        node delay = 20 msec

Route 5-1 = route 5-2 + route 2-4 + route 4-1 + 2 node delays
         = 71.1 + 0.05 + 35.15 + 40 = 146.30 ms
Route 5-4 = route 5-2 + route 2-4 + 1 node delay
         = 71.1 + 0.05 + 20 = 91.15 ms
Route 5-6 = Route 5-3 + route 3-6 + 1 node delay
         = 105.05 + 89.125 + 20 = 214.175 ms
```

Figure 7-3 Routing table.

4-2-1 and 4-1. The natural impulse is to reduce the graph, such as in this situation, replacing route 5-2-4-1 with route 5-2-1, which looks shorter; but queuing delay makes the latter route longer.

The graph shown will not change unless a new node is added to the network, but the path of least delay will change very rapidly. An example of such a dynamic change would be if the shortest path delay were calculated and then a file were transferred that completely filled the queue. The node would then have a large queue delay and all nodes connecting to it would be notified. They would immediately calculate the shortest delay, bypassing the node transferring the file. In adaptive routing tables, the path with the shortest delay is only a snapshot of the network for small intervals of time.

The routing tables between adjacent nodes are continually exchanged so that each node can continually update its own routing table. If this occurs too often, instabilities occur that cause oscillatory behavior. On the other hand, if data are not exchanged frequently enough, the dynamic behavior of the system makes the updated nodal data useless.

Another approach to adaptive directory routing is the use of a central controller that generates all routing tables. This unit can broadcast updates to all

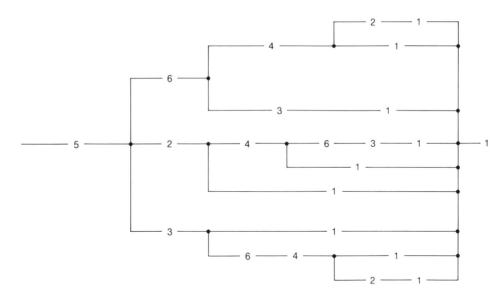

Figure 7–4 Tree diagram representation showing all possible routing paths between node 1 and 5 of Figure 7–2(a).

nodes on a special channel or periodically it can broadcast the data on the primary channel. One problem with this technique is that the network degrades drastically if the central controller fails. There are a multitude of routing algorithms; however, this brief introduction will give the reader some insight into what is involved in these algorithm designs.

Local-area Networks

Previous discussions have given an overview of various networking topologies and network schemes. In this section of Chapter 7, LANs will be investigated from primarily a network point of view. This layer of the LAN hierarchy is primarily a hardware-oriented layer. Let us examine this statement for a moment.

The hardware must be configured in some topology to form a network, whether it be a star, ring, or bus structure. Routing must be implemented, which can be a dedicated microprocessor, microcomputer, minicomputer, or the like. The routing algorithm is usually embedded in the network node or a central controller and rarely altered. This may be considered as part of the hardware with firmware embedded. This layer can be difficult to divide into software and hardware. As various networks are examined, the division between the two for that network will become apparent.

This section will address CSMA, CSMA/CD, and token-passing protocols with ring, star, bus, and other topologies.

Collision-sensing Networks

The topology of the collision-sensed network can be a ring architecture in which transmission may begin whenever a node does not detect a carrier. This is CSMA (carrier-sensed multiple access). Rings may also be designed with CSMA/CD (collision detect). The latter ring not only listens for a carrier but also has collision detection. The advantage is that when a collision occurs the node will know before it receives its own data back with errors. Bus topologies can also use this protocol for determining what the network status is before transmitting. The bus can be a broadband cable with FDM channels dedicated to CSMA or CSMA/CD protocols. For a star topology, collision-sensed networks are not very practical because each node has multiple transmission paths. However, the node itself may be a ring topology with multiple gateway stations that connect to other ring-type nodes. Thus star networks cannot be completely ruled out.

Ring CSMA

Ring structures have certain characteristics, which are part of their design philosophy; they are as follows:

1. Messages are passed in one direction from node to node.
2. Propagation delays are kept small; link distances are less than 10 km.
3. Nodes have an ordering inherent in network design (i.e., they are dependent on their physical placement on the ring).
4. The connectivity of the network is minimal. An N-node network will require $2N - 1$ connections.
5. To send a message to all nodes on the ring requires $N - 1$ relays of the message.
6. Each node in the ring is an active element (i.e., it amplifies signals, reads addresses of the packet, and processes the data for errors).
7. The ring limits the number of messages traveling around the ring at any one time.

Let us now investigate how these contention rings behave according to the design philosophy given.

Before a node begins to transmit, the receiver listens to the ring for traffic (i.e., it monitors the ring interfaces). If no data are present on the ring, the node begins transmitting. At the end of transmission, it may append a token to the last packet, which is the most efficient technique. When another node wishes to transmit, it listens for the token at the end of the packet. The token is changed to a flag (delimiter) by the node waiting to transmit. The waiting node then begins transmitting and inserts a token at the end of its transmission.

When data are received by the node that transmitted them, the data are removed from the ring along with the flag and token at the end if it is present. Thus during heavy traffic loads the contention ring behaves similarly to a token ring.

If the ring is at idle and two stations begin to transmit, a collision of the packets occurs. The receivers of the two transmitting nodes compare the transmitted data with that which is received and both nodes will detect errors as the data are taken off the ring. Note that the node can be listening, repeating, or removing data. There is another condition, node failure or power failure at the node; this type of failure would be catastrophic if no provisions were made for it. Diagrams of these connections are shown in Figure 5–7(b) for the electrically connected nodes (copper cable) and Figure 5–12 for an optical switch. Both switches are shown incorporated in a ring network in Figure 7–5(a) and (b). Within the interface circuits are embedded the necessary electronics, such as relay solenoids (not shown) and the necessary drive circuitry. The relay contact networks shown in Figure 7–5(a) are configured in the power-off state, which allows the nodes to bypass signals through the closed contacts. When the interface is under power, R, R^1 and T, T^1 connect the receiver and transmitter, respectively, to the twisted pair network. Note that the two closed contacts are wired in series, because the relays used for these circuits are equipped with four sets of transfer contacts (four form C's).

The other ring network [Figure 7–5(b)] is a fiber-optic version of the same ring network. This network is fiber-optic throughout to the transmitters and receivers. A 2 × 2 switch is provided as shown in the diagram; it will disconnect the transmitter and receiver of the node and form an optical bypass if a power failure occurs. The solenoid for this switch is again embedded in the interface circuitry and not shown here.

The fiber-optic version has many attributes that make it appealing for a ring network design solution, which will be explored here. Fiber-optic cable plants require no matching networks, and each node is electrically isolated from its neighbor except for the power mains. Bandwidths of the fiber-optic cable plant are in excess of 800 MHz-km for 50-μm core/125-μm clad fiber. For military and industrial applications, it may be procured with a measure of nuclear radiation hardness. The single-mode cable plants (single-mode fiber only supports the transmission of a single wavelength) are even more impressive. These waveguides have distance bandwidths of 10 GHz-km. The losses of standard 50/125-μm cable are 4 dB/km for 800-nm wavelength operation, 2 dB/km for 1300-nm wavelength, and 0.9 dB/km for some of the newer 1550-nm wavelength waveguides.

Fiber optics does have a few problems, which will eventually be overcome. Most multimode fiber-optic switches have relatively large insertion losses. The loss can be as large as 1.5 to 3 dB. Therefore, if a ring network had four node failures, losses would be in the 6- to 12-dB range; with the switches in a bypass

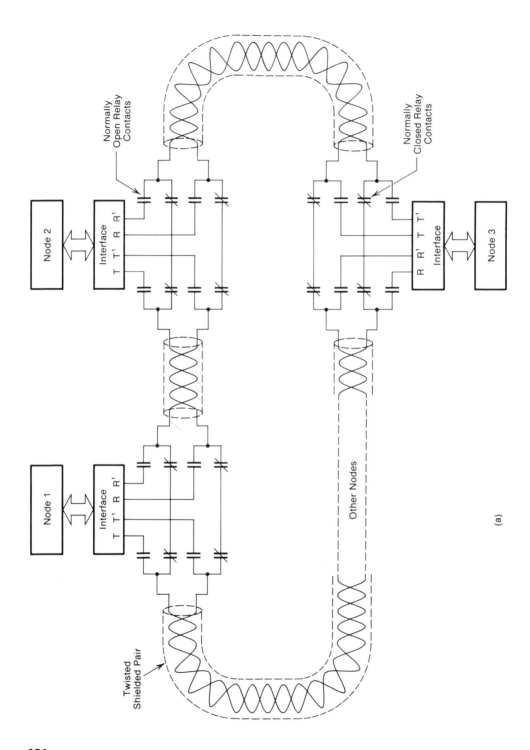

(a)

174

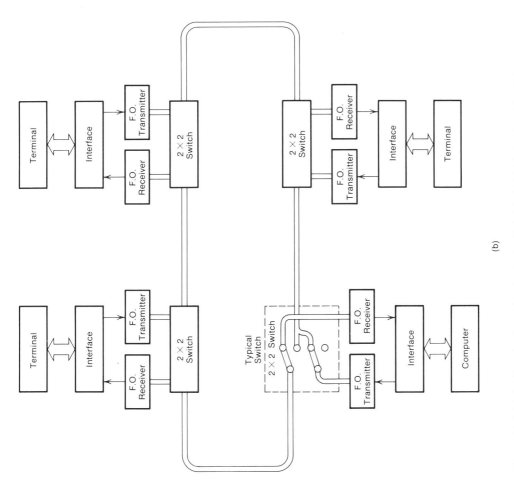

Figure 7-5 (a) An N node ring network wired with twisted shielded pair cable. Relay contact configurations are shown wired for fail-safe operation. (b) Four-node fiber-optic ring network with an optically wired fail-safe switch.

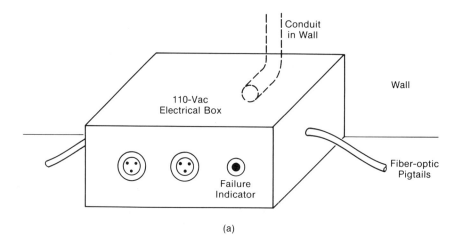

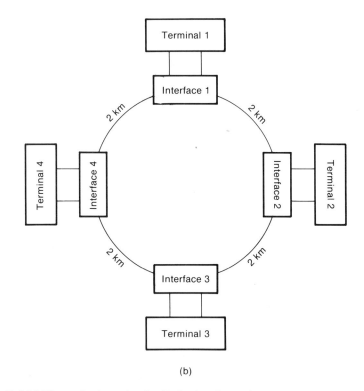

Figure 7–6 (a) Fiber-optic ring network with the interfaces shown connected to a terminal via external connections. (b) Mechanical detail that depicts a wall-mounted interface. (c) The optical and electrical circuit.

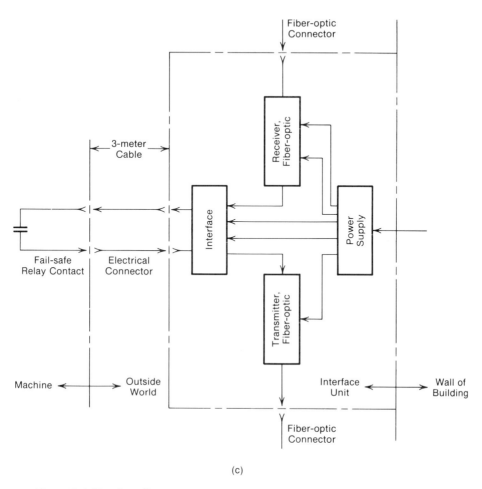

(c)

Figure 7-6 (Continued)

condition, soon losses would induce a network failure, a condition that they were to prevent. Single-mode fiber-optic technology will eventually solve this switch problem.

The cost of fiber-optic components is rather high, but as the number of applications increases, component costs will decrease. Fiber-optic switches may be removed from the design entirely through the scheme shown in Figures 7-6(a), (b), and (c). The ring is shown in Figure 7-6(a) with the interfaces wall mounted similar to electrical outlets, as depicted in Figure 7-6(b). These outlets always remain under power. With this design the outlets function as a repeater when a machine is removed or if the machine power is turned off. A block diagram of the wall-mounted outlet is shown in Figure 7-6(c).

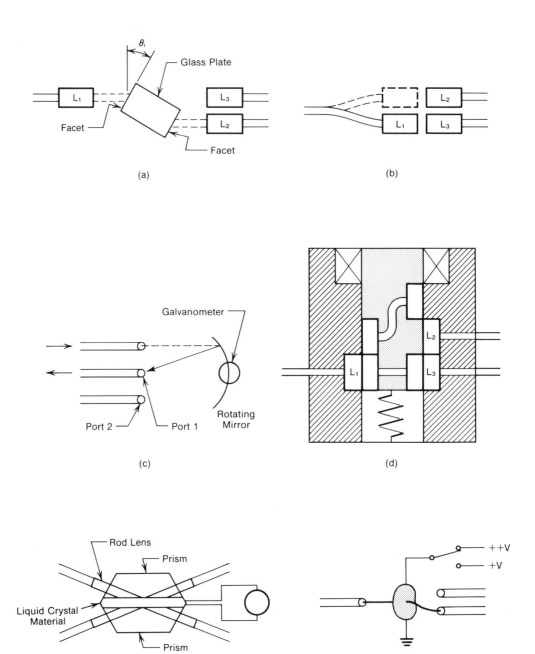

Figure 7–7 Optical switches. (a) Moving plate (glass) type. (b) Movable fiber switch. (c) Galvanometer switch. (d) Moving core switch. (e) Liquid crystal switch. (f) Crystal switch.

The block functions of Figure 7–6(c) are fairly straightforward. A power supply is provided in the outlet box to power the transmitter, receiver and interface circuitry. The interface circuit provides a number of functions. A fail-safe relay connects the fiber-optic receiver to the transmitter if the machine plugged into the outlet is removed. A failure of the power supply, transmitter, receiver, or any other electrical function must initiate a failure warning on the outlet. The outlet may be physically bypassed by a repairman and repaired while the network is operating. This can be accomplished if the fiber-optic connectors are disconnected from the ring and an optical jumper is installed. Patch cords may also be used with the connectors brought out to the face of the outlet. When a failure is noted by office personnel, a plug may be installed in the patch cord receptacles with a fiber-optic jumper embedded within it.

Some of the techniques for switch design are shown in Figure 7–7. Figure 7–7(a) depicts a movable glass plate used to divert the light beam from lens L_1 to L_2 or L_3 depending on its position. Lenses L_1, L_2, and L_3 expand the beam or contract the light beam (e.g., if the beam travels down the waveguide to L_1 where it is expanded). Then it either passes to L_3 or is diverted to L_2 if in the position shown in the diagram. Lens L_3 or L_2 demagnifies the expanded beam to the original size. The beam is expanded as it passes through the switch to reduce alignment tolerances. This type of switch exhibits losses of from 1.5 to 3 dB depending on how close tolerances can be held.

The switch depicted in Figure 7–7(b) diverts the beam using a technique similar to a relay contact. The lens L_1 and waveguide are movable, as shown in the figure; the solenoids for these first two switches are not shown, but one should be aware that these are electromechanical devices. Therefore, the switch actuation times are in the range of 1 to 15 ms depending on their mass, spring, and damping constants.

The switch in Figure 7–7(c) is a rotary switch that uses a mirror to divert the beam and a precision galvanometer to move the mirror. When a voltage is applied to the galvanometer, the rotor moves the mirror with the position proportional to voltage. This of course is an electromechanical positioning system, and the switch actuation time is slow. A variation of the galvanometer technique is to replace it with a stepping motor for positioning the mirror.

The last electromechanical switch is shown in Figure 7–7(d). This device is tailored after a hydraulic spool valve. When the switch is in the position shown in the diagram, the light beam from L_1 passes through the switch to L_3. If the solenoid is active, light will pass from lens L_1 to L_2, which diverts the beam. The actuator is electromechanical, as it was with the previous switches; therefore, reaction is again fairly slow.

The diagram shown in Figure 7–7(e) is an electrooptic device employing the use of liquid crystal technology. The construction consists of four waveguides coupled to a prism using rod lenses; the prisms have a thin layer of nematic liquid crystal between them (the layer is 6 to 8 μm thick). The operation is as follows:

When an external electric field is applied to liquid crystal material, the molecular order aligns. The light from the rod lens is totally or partially reflected at the liquid-to-glass interface. The amount of reflection is dependent on the control voltage applied to the liquid.

These switches have fast switching time, but insertion loss is in the neighborhood of 2 dB. Crosstalk is fairly low, approximately in the 40- to 50-dB range. Physical packaging is rather compact, about ½ by ½ by ¼ inch, which indicates they may be stacked for multiple-pole functions.

Switching time is dependent on electrical R and C parameters. This switch is electrostatic (i.e., very little current will flow, which is only leakage). One advantage is that no appreciable heat is generated as in some of the solenoid devices. The upper and lower temperature extremes of the device are dependent on the liquid crystal. Many of the materials have freezing points close to water.

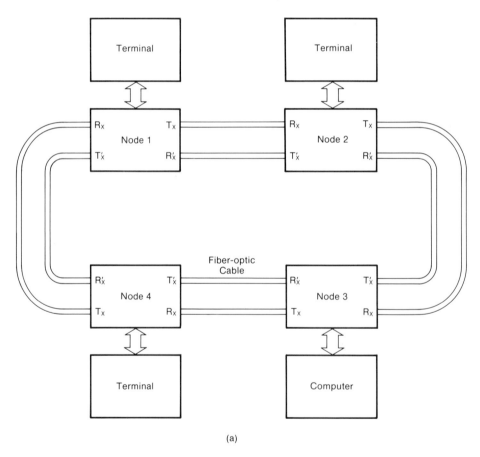

(a)

Figure 7-8 (a) Redundant counterrotating ring architecture. (b) Redundant counterrotating ring with a transmitter failure.

A final switching technique is shown in Figure 7–7(f). This is magneto-optical (diffraction of the light beam is controlled by a magnetic field). This method is quite often employed in thin-film switches for monomode waveguide. The technology is relatively new and has led to forms of hybrid-circuit signal processing. A great deal of R&D has been devoted to these methods because they are useful for fabricating integrated optic circuits.

Another technique used to improve reliability is the dual ring-architecture, which allows node failures to be bypassed. The bypass can be either electrical with relays or optical using fiber-optic switches. A block diagram of this ring is shown in Figure 7–8(a); this type of ring may also be implemented using copper transmission mediums such as copper coaxial cable or twisted pairs.

Let us examine the block diagram for a moment; this particular network is a dual redundant counterrotating ring. If a node transmitter failure occurs such as T_x of node 1, then the ring is reconfigured as shown in Figure 7–8(b). The pri-

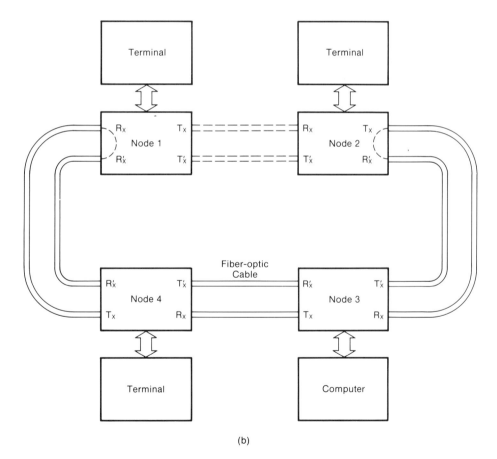

(b)

Figure 7–8 (Continued)

mary transmission path occurs through the T_x and R_x transceiver components. However, when a failure occurs, the secondary path R_x^1 and T_x^1 through the transceiver also becomes active. Note in Figure 7–8(b) that the ring will physically look as it did in Figure 7–8(a), but the cables shown in dashed lines will not be active; therefore, the actual path resembles Figure 7–8(b).

Some observations can be made as follows: If a second node failure occurs before any repairs are made to the ring in Figure 7–8(b), then the network will lose continuity (i.e., it will break into two isolated rings). These rings will function as two separate rings until a repair is made. Another item to consider is that the propagation delay through nodes and cable plant has essentially doubled; this will of course affect network performance.

Other observations that are not so obvious are the electrical and optical modes of failure that affect performance. As an example, if node 1 has a single power supply for T_x, T_x^1, R_x, and R_x^1, a catastrophic failure of the power supply will remove the node and its failure test circuitry. Nodes 2 and 4 may not have defective components, but they must communicate over the secondary ring and determine that node 1 has failed. This will cause the logic to become rather complex. The nodes must therefore have isolated primary ring, secondary ring, and logic supplies. In some cases where high reliability is necessary, a battery backup may be required for node interface circuitry.

For the situation where optical switches are used in the fail-safe mode, the reliability will increase at the expense of higher-cost components and higher insertion loss across switches. A series of references is provided for the reader interested in this type of architecture; see references [3] through [7]. Some forms of the dual redundant counterrotating rings have been successfully implemented. Examples are presented in the references.

CSMA/CD Bus Network: Ethernet

As an example, Ethernet has collision detection; that is, it is CSMA/CD. In Chapter 6 the protocol was discussed. Let us now examine some typical networks using Ethernet implements.

Figure 7–9 is an example of a typical Ethernet installation. The functional characteristics of the implements will be given prior to a discussion of the system aspects of the network. The definitions are as follows:

1. *Host computer:* This computer accepts Ethernet controller cards such as Ethernet-to-Q bus communication controller (DEQNA), Ethernet to UNIBUS communication controller (DEUNA), Ethernet-to-Professional 350 Communication Controller (DECNA), the Local Network Interconnect DELNI, and others to be discussed as they are presented.
2. *R (local repeater):* The local repeater allows the connection of two Ethernet cable segments (500 m each) that are no further than 100 m apart.

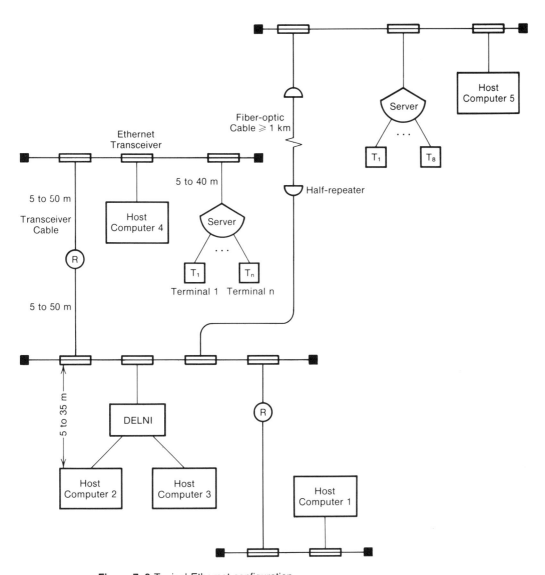

Figure 7–9 Typical Ethernet configuration.

3. *R (remote repeater):* Remote repeaters connect all cable segments that are farther than 100 m apart and less than 1000 m apart (these are fiber-optic links).

The remote and local repeaters are connected to the cable segments via two transceivers and transceiver cables. The remote repeater consists of two local repeaters each with a fiber-optic interface board. The local repeaters are also called half-repeaters because two are required to configure a remote repeater.

DELNI: Local Network Interconnect

The DELNI is a low-cost data concentrator that can be implemented as a stand-alone, hierarchical stand-alone, or network-connected unit. In the stand-alone configuration, it can interconnect a maximum of eight systems, and an Ethernet coaxial cable or transceiver is not required.

The hierarchical stand-alone DELNI also allows the interconnecting of these units to form a network without Ethernet coaxial cable and transceivers.

The network-connected DELNI allows the network to be expanded without adding cable segments. Each cable segment can then have over 100 implements connected because each DELNI expands the single transceiver connection to it by a factor of 8. A further discussion of DELNIs will be deferred until later when network examples are examined.

Transceivers

These devices were discussed previously in Chapter 5.

Servers

Four types of servers will be considered:

1. The *terminal server,* which connects multiple terminals to one transceiver. This technique expands the LAN capability without the addition of new cable segments.
2. The *router server,* which is a unit that transfers data packets between nodes on an Ethernet and other Ethernet LANs or remote nodes.
3. The *X.25 gateway* (router), which connects LANs to X.25 packet-switched data networks and to remote systems.
4. The *SNA gateway,* a server that connects Ethernet LANs to IBM host computers in an SNA network.

The units described are Digital Equipment Corporation (DEC) components; however, other manufacturers make Ethernet components, such as Interlan, Inc. Before procurement of any components, it is necessary to examine the wares of several manufacturers to obtain the best price and also to ensure that they are not Ethernetlike components (i.e., the components are 100 percent compatible).

In Figure 7–9, note that four coaxial Ethernet cable segments are connected via repeaters. The three connected by local repeaters could be interconnected within a building such as a college campus administration building. The distances between all the implements must be kept within close proximity. For the fiber-optic repeater case, the segments may be located in another building, such as a laboratory. This allows user terminals T_1 through T_8 on the server to access not only the local host computer but any other hosts within the network. Sharing of these expensive resources allows more efficient use of them. If, for example, host 5 fails, then the other computers would allow the terminals on that segment access. Performance of the network would suffer, but a complete failure would not occur.

Another example will be considered before moving on to another topic. Ethernet is a one of the most widely implemented technologies; therefore, it should be examined more thoroughly by the use of network examples.

Figure 7–10 is a block diagram of a typical factory implemented with Ethernet. Engineering is located at another building less than 1 km away. Note that the sales offices are connected to the factory via normal telephone switch network facilities. For larger installations, such as General Motors or IBM, all the factories may be connected to an installation that is a communications facility or private telephone exchange. The corporate offices may also be located in another building and require a remote link using fiber optics, telephone facilities, or satellite as the transmission medium. The decision depends on economic issues, distance between buildings, transmission rate, and other factors.

The factory shown in Figure 7–10 provides automatic assembly and test devices. For example, assembly may be accomplished by robotics, as well as testing. Data-entry stations located throughout the factory provide timely information on the entire manufacturing process. The exchange of data between manufacturing and corporate offices is accomplished through the router server.

The laboratory may require certain types of data from manufacturing, such as statistical information, or engineering modifications may be necessary to manufacture products. The fiber-optic link allows two-way communications between the two data bases. Continuous monitoring of laboratory testing will keep the scientists and engineers informed of the progress of tests. Failures can be quickly acted on, thus saving valuable engineering and research time.

As a final note, an Ethernet transceiver is shown in Figure 7–11, which will allow the reader to become acquainted with the actual hardware. This transceiver package is connected to the station hardware via a transceiver cable. A vampire clamp on the transceiver pierces the outer coaxial packet and makes contact with the inner conductor, and contact is made with the outer shield; this completes the tap.

Token-passing Networks

Token passing is used on rings, baseband bus, and broadband bus topologies. It allows a more orderly transfer of data and an upper bound on waiting time between network access by each station. The networks remain efficient under heavy loading due to their collision-free operation. Token networks are useful for applications requiring guaranteed network access time. Examples of such requirements are real-time processing, process control, factory automation, and packet voice transmission.

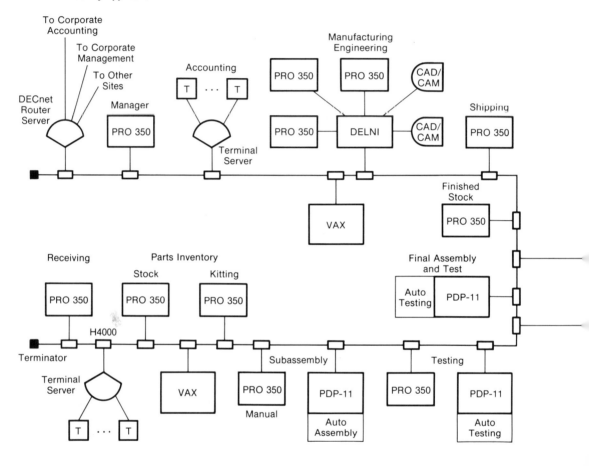

Figure 7–10 A typical factory implementation with Ethernet, with the engineering building remotely located.

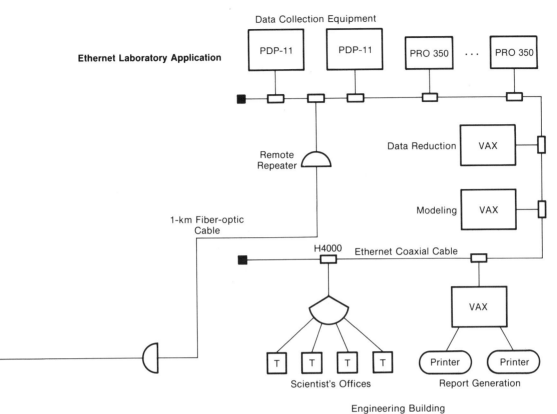
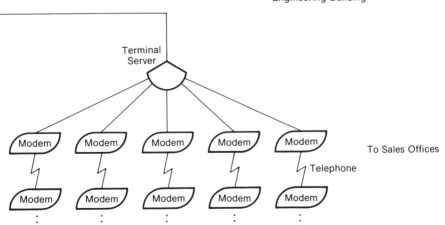

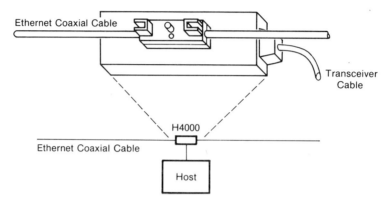

Figure 7-11 Ethernet transceiver.

Token passing is not without its penalties; otherwise, CSMA/CD would become obsolete, which of course is not the case. Token-passing techniques are sensitive to the number of nodes in the network. Excessive bandwidth overhead due to passing the token to idle nodes causes this condition, especially when network loads are light. These networks are also sensitive to the physical length of the ring or bus structure, which of course increases propagation delay.

Token Rings

One leading manufacturer using token rings is IBM. The rings are local loops connected to a central distribution panel. These local loops are connected to workstation terminals or host-computer nodes as shown in Figure 7–12. This is a ring-

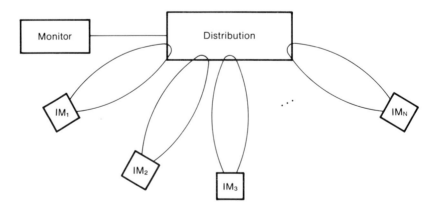

Figure 7-12 Implements on an IBM token ring.

star configuration. Each of the rings operates at a 4 Mbit/s transmission rate, and the data are Manchester encoded. The delimiters for the packets (same as flags) are formed by Manchester code violations. The packets have a 4-byte destination address, as well as a 4-byte source address. The first two bytes identify the destination on the ring. A control byte is provided to maintain ring integrity; it indicates whether the traffic is synchronous or asynchronous.

The ring monitor occasionally declares a period of only synchronous operation. This monitor can refuse new synchronous/asynchronous requests until sufficient bandwidth is available. During synchronous operation, the monitor permits only single packet transfers between the call originator and the answering station (i.e., a single packet is transmitted and a reply will be in the return packet). Synchronous operation can be used to implement both telephony and non-IBM devices. IBM and Texas Instruments have produced controller ICs to emulate the token-passing ring protocol.

Token Bus

During normal operation, the token is passed in a direction toward decreasing node addresses. When the lowest address has been reached, the token is passed to the highest. This is a form of logical ring, even though the network is a bus structure. As is the case with all token networks, nodes may only transmit if they have the token. Otherwise, the node may function as a repeater with error-checking capability.

A station can remove itself from the bus by using a *set next node* frame (see Figure 7–13). The station must hold the token to perform this task. First, the station informs its predecessor of the removal; then the successor is informed on subsequent cycles. After these two functions are performed, the node becomes transparent to the network.

00000000	Destination Address	Source Address	Address of New Successor

Figure 7–13 Set next node frame.

The existence of a logical ring creates a need for maintenance. The routine functions are as follows:

1. Reconfiguration of the logical ring.
2. Adjustment to and maintenance of the algorithm (e.g., maximum time a station may hold the token, larger address field, new control functions defined).
3. New stations being implemented on the ring.

Certain stations may be designated to perform these maintenance functions.

The first two are performed in a straightforward manner, while the third is not quite so simple because of new station recognition. The process of recognizing a new station on the ring is performed using controlled contention, which is the same process used for a lost or damaged token.

Each station designated as a ring maintenance station initiates controlled contention each nth time it receives the token, where n is set by the system. The maintenance station transmits a solicit successor frame to start the process, as shown in Figure 7–14. Maintenance stations with the lowest address require two windows. Stations that are to join the network will place set next node frames in the appropriate window.

| 00000000 | Destination Address | Source Address | FCS | Window 1 | Window 2 |

Figure 7–14 Solicit successor frame.

When a token is lost or damaged, a period of silence on the bus indicates this condition. The maintenance stations then begin the contention process after the period of silence, and the logical ring will be reestablished.

If the maintenance station hears a response after the last token procedure, it adds the new station to the logical ring. If there is no response, the station concludes the maintenance phase. If a collision occurs on the system, a resolve contention frame follows with four windows transmitted. Each demand station selects one of the window slots according to the first two bits of its address and transmits a set next node frame. If a collision occurs during this process, the station drops its demand.

If the controlling station detects a second collision, the resolve contention frame is repeated, and the demander nodes respond using the windows with the third and fourth bits of their addresses. This procedure continues until the controlling station receives a valid set next node, there are no further demanders, or the maximum timeout is exceeded (this prevents thrashing about when damage has occurred to the network.)

Transfer of data is a rather straightforward process. A station may not transmit unless it passes the token, which it must capture as the explicit token frame passes through the node. After the token is captured by the node, it can transfer data for a limited amount of time determined by a timer that is set at the time of token capture.

Frames to be transmitted have the following prioritized classes of service:

1. Synchronous
2. Urgent asynchronous
3. Asynchronous
4. Time available buses

Network Layer

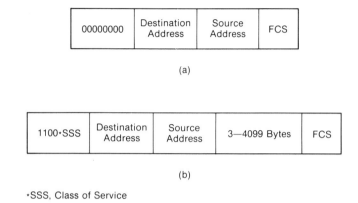

Figure 7-15 (a) Token frame. (b) Data transfer frame.

Frames on the bus are cleared according to their priority.

A description of the explicit token frame and data transfer frame is given in Figure 7-15.

The attributes of the token bus are similar to those of ring topology; that is, performance is rather good for lightly loaded networks, for networks that have small numbers of nodes, and for short cable spans. Performance will degrade with increased data density, but transfer of data will remain orderly and reliable.

When transit delay is critical, such as in telephone system voice recognition and certain types of encryption, token passing is unsuitable. Token passing has an upper bound on transit delay, which is dependent on the number of stations (nodes). Token bus algorithms are extremely complex and were very costly to implement. However, with the advent of the new microprocessor technology and low-cost memory, token bus technology will become more cost effective and appealing to network controller designers.

Microprocessors such as the MC 68000 series (Motorola), 32032 (National Semiconductor), and IAPX 432 (Intel), which have the computing power of minicomputers, can be used for the token passing bus algorithms to easily implement them. At the time this book was being written, the cost of these devices was prohibitive; but they will decrease in cost dramatically as demand increases and manufacturing processes improve.

Star Networks

Thus far, only a brief introduction has addressed star networks. They are a rather important class of networks because they lend themselves well to fiber-optics technology, which will eventually become a formidable medium for networking.

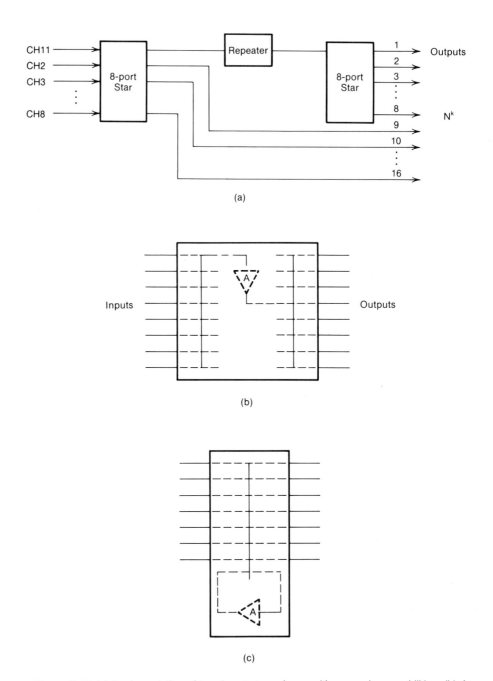

Figure 7–16 (a) Implementation of two 8-port stars shown with expansion capabilities. (b) A form of active star in which the repeater amplifier is embedded in the unit. (c) Implementation of an active star with one input and one output internally connected to produce an output signal equal in magnitude to the original signal. (d) The increasing complexity of star networks.

Star-configured local-area networks are rather easy to implement in fiber optics because star couplers are simple devices to construct and input data are broadcast to all output parts. Let us now examine some examples of star networks. The investigation will be limited to fiber-optics and hybrid star networks. The latter star is a combination of electrical and fiber-optic design.

Figure 7–16(a) is an example of a two-node, eight-port star (i.e., eight inputs and eight outputs). A repeater is added because the output will be attenuated by $1/n$ because the signal is distributed to the other outputs and no gain elements are present. Other losses include coupling losses due to connectors or splices and excess loss, which is attributed loss not accounted for, such as radiation loss. To restore the signal level to its original state, a repeater with an amplifier is necessary. Then outputs 1 through 8 will be equal to outputs 9 through 16. Figure 7–16(b) depicts a star with the amplifier included (commonly referred to as an active star).

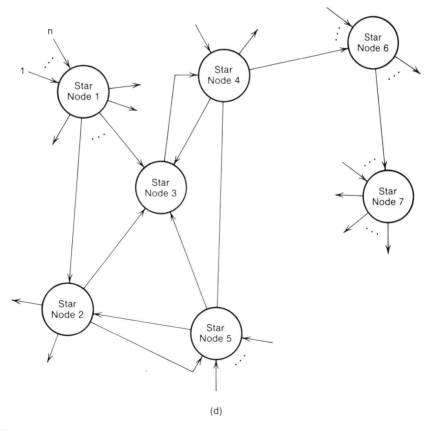

(d)

Figure 7–16 (Continued)

Another technique for reinforcing the star output levels is another form of active star shown in Figure 7–16(c). One of the output ports is passed through a repeater amplifier; the signal is reinjected into the star at a level sufficient to produce outputs that are equivalent to the original input. Active stars all require external power supplies, which are a source of network failure. For situations where active stars can be removed as an implementation candidate, by all means remove them. The active star will no doubt be more expensive.

Before going on to Figure 7–16(d), let us examine further the fiber-optic star. The receiver will have a signal-to-noise ratio determined by Equation 7–1.

$$\text{SNR} = 10 \log \left[\frac{i_s^2}{\Sigma\, i_{optical}^2 + \Sigma\, i_{electrical}^2} \right] \quad (7\text{–}1)$$

$$i_{optical}^2 = i_q^2 + i_g^2 + i_B^2 \quad (7\text{–}2)$$

$$i_{electrical}^2 = i_T^2 + i_d^2 + i_L^2 + i_F^2 \quad (7\text{–}3)$$

The signal parameters are defined as follows:

$$i_s^2 = \text{signal level (mean square value)}$$

$$= \frac{1}{2}\left[\frac{\eta e}{h\nu} G m P_0\right]^2$$

where $\dfrac{\eta e}{h\nu} = r$, responsivity of the detector

G = avalanche gain
η = quantum efficiency
m = modulation index
e = electron charge
P_0 = optical power at the detector

Therefore,

$$i_s^2 = \frac{1}{2}[rGmP_0]^2 \quad (7\text{–}4)$$

Let us now examine the noise parameters attributed to optical parameters, Equation 7–2.

$$i_q^2 = \text{quantum noise}, \quad B_n = \text{noise bandwidth}$$

$$i_q^2 = \frac{2e^2\eta}{h\nu} P_0 G^2 F_d B_n = 2r e P_0 G^2 F_d B_n \quad (7\text{–}5)$$

where F_d is the noise figure due to the random nature of the avalanche process.

i_g^2 = incoherent background radiation noise components.

$$= \frac{2e^2\eta}{h\nu}P_G G^2 F_d B_n = 2erP_G G^2 F_d B_n \qquad (7\text{-}6)$$

The third term in Equation 7–2 is the LED beat noise; this term does not appear in lasers.

$$i_B^2 = 2\left(\frac{eG\eta P_0}{h\nu}\right)^2 \frac{B_n}{N_m \lambda_w}\left(1 - \frac{B_n}{2\lambda_w}\right) \qquad (7\text{-}7)$$

where λ_w is the spectral width of the source and N_m the number of spatial modes received.

The electrical parameters of Equation 7–3 are as follows:

i_T^2 = thermal noise currents

$$= \frac{4KTB_n F_t}{R_{equiv}} \qquad (7\text{-}8)$$

i_d^2 = noise due to dark current

$$= 2eI_d G^2 F_d B_n \qquad (7\text{-}9)$$

i_L^2 = leakage current component of noise

$$= 2eI_L B_n$$

i_F^2 = channel noise (FETs only) $\qquad (7\text{-}10)$

$$= \frac{3.73\,(\pi c)^2 B_n^3 KT}{g_m}$$

$$\beta P_0 = \frac{1}{n_s}\alpha_c L_e P$$

where L_e = excess loss
α_c = coupling loss
n_s = number of outputs

Then

$$i_q^2 = 2reG^2 F_d B_n \sum_{k=1}^{n_s} \beta_k P_k \qquad (7\text{-}11)$$

For a star coupler, the $i2_q$ term is modified to reflect the star inputs. Each star input is represented by Equation 7–11.

The SNR general equation for the star is Equation 7–12; this equation was derived substituting Equations 7–4, 7–5, 7–7, 7–8, 7–9, 7–10, and 7–11 into 7–1.

Let $X_{FO} = 2reG^2B_n$

$$\text{SNR} = 10 \log \left\{ \frac{1/2[rGmP_0]^2}{X_{FO}F_d \sum_{k=1}^{n_s} \beta_k P_k + X_{FO}F_d + X_{FO}P_0^2 \frac{r}{eN_m\lambda_w}\left(1 - \frac{B_n}{2\lambda_w}\right)} \right.$$

$$\left. + \frac{4KTB_nF_t}{R_{equiv}} + 2eI_dG^2F_dB_n + 2eI_LB_n + \frac{3.73(\pi c)^2B_n^3KT}{g_m} \right\} \quad (7\text{--}12)$$

Equation 7–12 is a rather formitable equation describing SNR for the star at the fiber-optic receiver. For an SNR approximation, often only the i_q^2 and i_T^2 terms are considered. From experience with various star networks, the author has noted that the SNR is constant for a star network if the number of inputs is less than eight. This is due to the small i_q^2 noise component for each input port. When very low noise receiver amplifiers and detectors are used, the quantum noise causes a reduction of SNR with fewer star inputs present or higher power at the inputs. This conclusion can be drawn from an examination of Equation 7–11, where the r, e, GF_d, B_n, and B_k are all constant, and only P_0 the fiber-optic power can be varied.

Inflections in a plot of measured values of SNR indicate an emerging dominance of a noise component. When observing measurements, multiple inflections show multiple dominance of noise components as certain threshold values are surpassed.

To optimize avalanche gain is rather involved with mathematical manipulations because F_t is a function G. Therefore, the usual method of taking the derivative and setting it equal to zero is not an easy task. A treatment of these gain-optimization techniques is given in reference [8].

The objective of this section is to make the reader aware of some of the noise components that affect the star performance. Only source partition and modal noise are neglected. Modal noise becomes prevalent for laser sources, which is of particular concern when designing video systems. This noise is statistical in nature and difficult to describe mathematically.

Modal noise is due to undesired modulation of guided light intensity arising from multipath effects in multimode waveguides. If the light from a waveguide is projected onto a screen, a speckle pattern can be observed that randomly fluctuates with time. The conditions that induce modal noise are as follows:

1. A source spectrum sufficiently narrow (i.e., a coherence time sufficiently long) to permit light guided in different modes to interface at the waveguide output plane.
2. Some form of spatial filtering at the output plane, such as would occur with connector misalignment, splices, couplers, splitters, or waveguide distortion in general.
3. Source wavelength shifts, movement of the waveguide, or both.

Partition noise is due to individual wavelength intensity variations with constant total spectral intensity. As multimode lasers are modulated, the intensity of each wavelength can fluctuate, redistributing its energy to other modes. Partition noise prerequisites are as follows:

1. Lasers must have (longitudinal) emission.
2. Lasers must have individual mode amplitude fluctuations.
3. The losses within the transmission medium must be wavelength dependent.

For an approximation of the total optical noise, the calculations are as follows:

$$\text{SNR} \cong 57 \text{ dB for a typical laser}$$
$$P_P = 1\text{-mW, typical}$$
$$P_n = 2\text{-}\mu\text{W noise due to optical components only}$$

An example of the SNR for the star is given in Problems 7–4 and 7–5.

Figure 7–16(d) is an example of a star-connected network that can grow to a very large complex network, which would soon become unwieldy and difficult to control. Thus when working with star networks, one must configure the architecture to allow for an orderly method of future expansion.

Hybrid Star Network

A hybrid fiber-optic star coupler in block diagram form is depicted in Figure 7–17(a). The input ports can be configured with receivers connected to the outside environment through pigtails and bulkhead connectors. These receivers may have decoding logic design into the receiver itself, or they may simply be bit drivers that convert the incoming optical signals to electrical. In this latter situation, a gate array may furnish the necessary logic to perform the decoding. Note, in the diagram, that data and clock from each receiver enter the serial-to-parallel interface. The assumption made here is that the incoming optical waveforms are self-clocking.

Serial-to-parallel conversion is accomplished with a series of USRTs (universal synchronous receiver transmitters) or a standard shift register. The USRT requires the data to be in a particular format. For example, the 6852 SSD, a Motorola device, has onboard buffering and operates in a full-duplex manner (i.e., it handles both serial input and output simultaneously). The data are processed in the USRT in 8-bit bytes, and sync characters must be included in the data. This type of serial-to-parallel conversion does not lend itself to packet formats. The standard shift register with buffers is more advantageous to use in this application.

Fiber-optic transmitters require serial data; therefore, a parallel-to-serial conversion register or USRT is again required depending on data format. Encoders may be embedded within the fiber-optic transmitters, designed in gate arrays, or constructed of discrete components.

The microprocessor is the heart of the system; it can be used to perform some of the following functions.

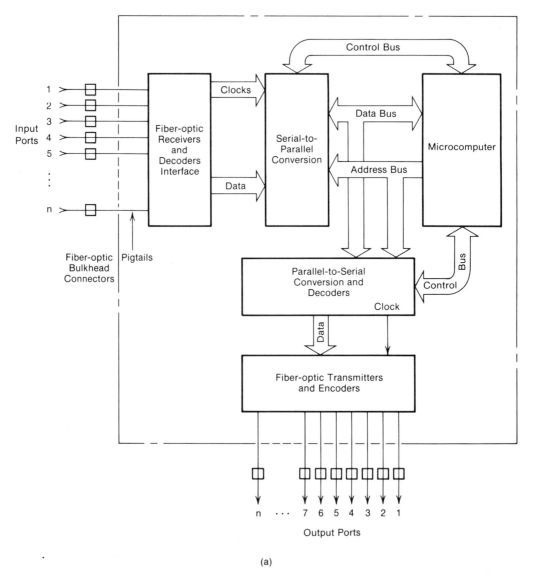

Figure 7-17 (a) Block diagram of a hybrid star. (b) Flow chart showing the functionality of a hybrid star.

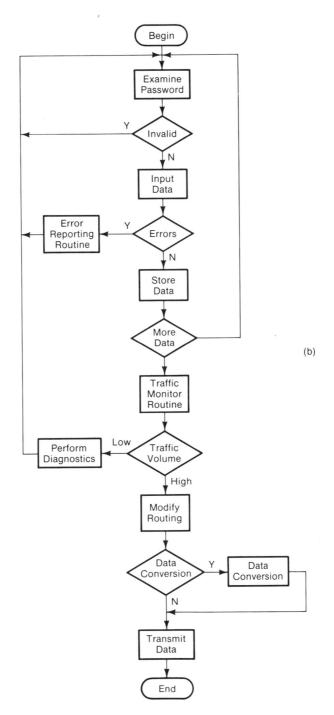

Figure 7-17 (Continued)

1. Error checking.
2. Log errors for maintenance purposes.
3. Monitor traffic.
4. Dynamic routing.
5. Initiate diagnostics during low traffic volume, such as local and far-end loopbacks.
6. Report abnormalities to a central maintenance node, which can be a minicomputer.
7. Data conversion (i.e., from one format to another).
8. Maintain a directory service (i.e., convert names to address locations or vice versa).
9. Security checks on passwords (e.g., check their validity).

The list of functions shown here can be much larger if digital voice is also processed.

For military applications, such as when networks fail due to damage, a simulation program may be embedded to remove nodes and allow the star couplers to circumvent the damage. This simulation would allow the network subscribers to access the network with a behavior during war-time situations.

A flow chart showing how the network functions is shown in Figure 7-17(b). This diagram is highly simplified; if the coupler flow chart were considered it would be quite involved and beyond the scope of the book.

The hybrid star is not without disadvantages, which require a trade-off analysis. Microprocessors are rather slow compared to the serial transmission rates of many LANs, which implies that processing time must be kept small or higher-speed devices must be used. For LANs in the 1-Mbit/s range, microcomputers similar to the Intel 8051 may be used. For higher speeds, bit-slice technology or discrete components must be considered. However, GaAs technology is producing some very high-speed devices. Eventually, when processes become efficient, cost will allow the implementation of high-speed hybrid star couplers. As an example of GaAs products, one manufacturer is producing a 900-MHz bit counter with 40 mW of dissipation. The same manufacturer will produce memories, shift registers, multipliers, dividers, prescalers, and multiplexer/demultiplexers.

Often, high-speed devices may be controlled by low-speed controllers. As an example, if a shift register can process 16 bit words, the input data can be strobed into the microprocessor every 1.6 μs for a 10-Mbit data rate. Therefore, with some type of direct memory access (DMA), data may be stored in a large memory, which can serve as a queue. The point of this discussion is not to design a hybrid star but to alert the reader to the design variations that exist for these devices.

One serious problem with these types of stars is delay. Any processing takes a finite amount of time to execute; therefore, delay at the star nodes occurs in proportion to algorithm complexity. Another is that if the hybrid star is not pro-

Network Layer

duced in large quantities, it cannot become very competitive with other techniques, such as fiber-optic stars or all-electrical star networks.

Gateways

Gateways are of prime importance because they are the glue that allows networks that do not have identical protocols or physical plants to be interconnected. As a simple example of a gateway, Ethernet data may be passed to another location with an Ethernet LAN via a telephone line.

Let us examine what is involved to perform this task. The Ethernet transmission rate is 10 Mbits/s and the telephone line transmission rate for high-speed modems is 9600 bits/s; an approximate 10:1 disparity exists between the transmission rates of the two networks. Therefore, when Ethernet-to-telephone network transmission occurs, a large buffer must store the data. For the reverse situation (i.e., telephone facilities to Ethernet transmission), a small buffer is needed to assemble the data in packet form before transmission begins. Ethernet protocol is not even remotely similar to the telephone facilities; therefore, a conversion process is necessary to accept either protocol and make them compatible.

Physical plant facilities for Ethernet are vastly different from their telephone counterpart. Electrical compatibility is necessary. Thus the task of producing an adequate gateway is quite formidable. Entire books can be written on this particular subject because of the large numbers of LANs available in the marketplace.

Often a manufacturer's network connects several implements and to redesign the system would be very costly. Gateways allow the locally connected equipment to communicate with a foreign network.

For connecting two site networks, a bridge using leased lines or other communications facilities is necessary. Each bridge end point requires protocol conversion of some type to allow end-to-end communication. The OSI transport layer and those above will pass packets unchanged and uninterpreted across the bridge network between the two site locations. The facilities of the intervening network are generally irrelevant and not available to the bridge user. However, in some circumstances the user of the network may require access to the network facilities. The following is a partial list of these circumstances:

1. Public and private packet-switched networks that support X.25, X.28, and X.75 interfaces (CCITT standards).
2. Private switched networks supporting X.21 interfaces (CCITT standard).
3. Private LANs, such as SNA, Ethernet, and Wangnet.
4. Video text networks (private and public).
5. Modem interfaces.
6. RS 232, 422, and 423 interfaces (EIA standards).

Gateways establish a virtual circuit between networks and transfer data across this circuit, but control, log in, and security are not implemented in the gateway. Its functions are restricted to the ISO network transport layer and in some cases the session layer. It will be difficult to design gateways to interconnect new LAN designs with older networks that do not have the ISO layer approaches.

Datagram versus virtual circuit service provided by gateways should be addressed before considering other issues. Datagram service through gateways is simpler to provide than virtual circuit service. Datagrams can simply be encapsulated with an envelope as it enters the network gateway, and then the envelope is simply stripped off when exiting. For long-haul networks, virtual circuits are usually employed because they ensure orderly delivery of message traffic. The IO X.75 protocol is appropriate for this situation.

The next issue to consider is addressing. Let us examine for a moment what is involved in passing a message across a gateway (see Figure 7–18). Before data can be transported through the transfer medium, the destination and the path between originating node and destination node must be known. Flow control is also the concern of gateways. The destination internetworking and origination networks may all have different data-transmission rates, which is the usual case. Network congestion may warrant gateway rejection or in some cases destruction of messages.

Network 1 Interface Protocol	Standard Internet-Working Interface Protocol	Transfer Medium	Standard Internet-Working Interface Protocol	Network 2 Interface Protocol

Figure 7–18 Gateway message transfer.

Other issues of importance are message fragmentation, privacy, and security. Fragmentation of message traffic occurs when internet, originating, and destination networks have different packet sizes. For situations where origination and destination networks are secure, the gateways will also be required to employ encryption.

Gateways add overhead to the LANs, which must be minimized if they are to be effective. Throughput optimization can be accomplished through gateways if they are made modular expandable. As an example of this architecture, let a single gateway module be capable of processing 1000 packets per second; and further suppose the design requirements for a particular gateway are 10,000 packets per second. But the original design only need be expanded using modules to meet the necessary design criteria. This allows the gateway designer a great deal of flexibility.

A useful technique to reduce loads on large networks is the use of dedicated subnetworks. Throughput increases because packets not intended for other LANs

Network Layer

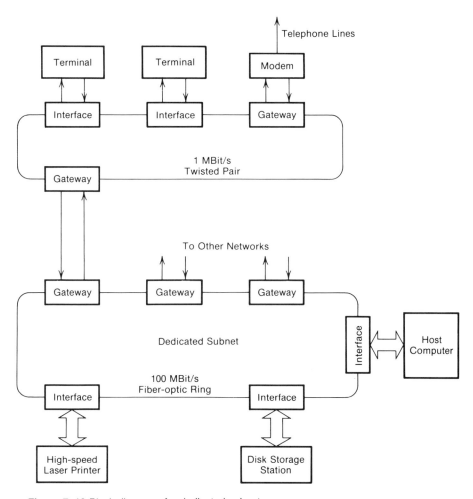

Figure 7–19 Block diagram of a dedicated subnet.

do not pass through the major gateways. As computers and other devices expand processing power and diversity, dedicated subnetworks become even more important.

As an example of a dedicated subnet, see Figure 7–19. The high-speed 100-Mbit/s fiber-optic link services the high-speed devices, host computer, laser printer, and disk storage station. The terminals and telephone gateway should not be serviced by the high-speed dedicated subnet. Due to their slow processing speed, they would reduce the throughput of the high-speed link. The gateways serve to match the speed of the two LANs. Other gateways provide access by other low-speed rings to the dedicated high-speed network.

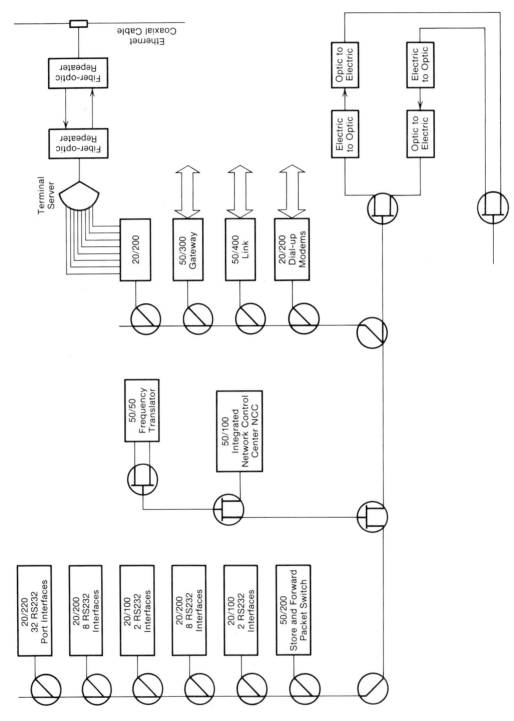

Figure 7-20 Typical Sytek broadband cable plant.

Further examination of the figure reveals that telephone gateways must further step down or up the data-transmission rate. This would present a problem if it were added to the high-speed link (i.e., the approximate 10,000:1 difference in transmission speed).

Broadband Cable Networks

A broadband network will be examined based on Sytek components. A block diagram of the network is shown in Figure 7–20. This network has features that make it unique from other networks, the fiber-optic links. Removing these components would make the network typical. The fiber optics were added to illustrate how one might use the technology to expand the networking capability of broadband cable. In certain instances, fiber-optic cable has been installed between buildings within a campus; thus, with careful planning, the existing cable plants may be used.

Each component shown in the diagram is especially designed to operate with broadband cable. The LocalNet 20/100 has been previously discussed in detail in Chapters 5 and 6. It provides a frequence agile modem that can software select one of 20 channels within a 6-MHz bandwidth for both transmit and receive. These units provide two RS232 or V.24 ports to all implements to be connected to the network, which are compatible with these standards. The LocalNet 20/200 and 20/220 are similar to the LocalNet 20/100 package with 4 or 16 packet communication units (PCUs) in a single enclosure. The 20/200 allows the user to connect eight devices to the network, while the 20/220 allows 16.

The LocalNet 50/200 is a microprocessor-based store and forward packet switch that logically connects up to 8 of the LocalNet's 20 channels. Multiple units may be cascaded or paralled to achieve any number of interchannels, with the additional option of providing redundancy.

The LocalNet 50/50 is a frequency converter and retransmission unit. This unit converts 10- to 106-MHz low-band signals to 196- to 262-MHz high-band signals and retransmits them in an outbound direction. These units are located at the network head end, and a failure of this unit causes a catastrophic failure of the network. A LocalNet 50/55 provides a backup unit mounted in the same package that will switch automatically, or the LocalNet 50/60 manually switched unit can be implemented.

A useful unit that can improve network performance is the LocalNet 50/100. This is referred to in the literature as a network control center (NCC). This device provides a centralized collection and display of network performance information. The NCC functions as an alarm center for up to five LocalNet channels. The NCC can also provide encryption, which provides a measure of security for the system. These control centers may be controlled from any terminal in the system.

The hardware description of the NCC is as follows:

1. It is equipped with a Motorola 8-MHz, MC68000 central processing unit.
2. A memory management unit (MMU) is provided.
3. The unit comes equipped with direct memory access (DMA).
4. Memory consists of 512 Kbytes of RAM with a dynamic parity check.
5. The NCC can be configured with either one or two 10-Mbyte Winchester disk drives.
6. These units also have 8 or 16 RS232 ports, which allows the implementation of other hosts.
7. The operating system of the NCC is a subset of the UNIX/V7 operating system.
8. A 20-MByte streaming cartridge tape drive is also provided.

This controller can be used for the following functions:

1. A data dictionary that records administrative and operational data for each network component.
2. A screen-oriented NCC data-base interface allows the network manager to validate network implement information, define their configuration and symbolic destination names (i.e., directory service), and define and validate the user passwords.
3. Component configurations can be initialized, monitored, and enforced with a special onboard algorithm.
4. The PCUs can be configured to automatically call the onboard name–server/access–controller ports of the NCC. This software translates a destination name to an address (directory), validates access rights (password validation), and creates a session between the originator and destination nodes.
5. Status information includes a display of channel throughput, error rates, and alarms.

Thus, the network communication controller provides the network with a great deal of flexibility.

The next item to consider is the transmission path between the Sytek network and Ethernet LAN. The terminal server provides eight terminal RS232 channels that can be connected to the eight ports of the 20/200. This technique allows network implementation with standard Ethernet components. The fiber-optic link can be up to 1 km in length; therefore, this system would be useful between campus buildings. One problem that may occur with the Ethernet bridge is the time delay budget. Time delay budget violations cause excessive collisions because the long delays make the cable appear not to be in use. A station desiring to transmit will then begin to transmit, and a packet can appear and cause a collision because the timeout before transmission is shorter than the actual delay in the network. Increasing the timeout (slot time) can correct this problem; however, this is a rather difficult task. An NCC may be required rather than a 20/200, which can be used to store and forward data. This technique makes the station appear as eight terminals communicating with the Ethernet LAN. This can ease the budget violation problems.

Network Layer

The second fiber-optic link in Figure 7–20 is shown as a method of extending the length of cable plants with fiber-optic technology. These electric-to-optic and optic-to-electric conversions are analog because the cable plant uses FDM for multiplexing data, video, radio, and so on.

Figure 7–20 does not show the implements on the network. However, the implements are standard devices such as printers, host computers, facsimile machines, telephone modems, and terminals. Adding the implements would only add confusion to the diagram. Any device to be connected to the bus must have an RS232 interface or it must have an interface that will emulate RS232 operation.

The gateway (50/300) shown in Figure 7–20 allows multiple foreign installations to be interconnected. This gateway also provides X.25 service, which makes the network quite flexible. For the reader contemplating a broadband cable installation, it would be advisable to shop around because of rapidly changing LAN technology.

The link shown in Figure 7–20 provides interconnection to other LocalNet services and facilities. This allows for future installations to be interconnected (future growth).

Broadband Fiber-optic Networks

At present, most of the devices discussed here are available, but not cost effective, or they are in the laboratory stage. Let us examine the cable plant of Figure 7–21. It is constructed with a single-mode waveguide optimized at 1550 nm with an attenuation of 0.2 dB/km (this is a reasonable assumption; the theoretical limit is 0.17 dB/km). However, it is not the cable and waveguide that present problems with these bus systems, but couplers due to connector and excess loss.

An assumption will be made that the minimum sensitivity for the up converter receiver is -42 dBm with a dynamic range of 24 dB. Node 1 is 1 kilometer from the up converter; the tap ratio of T_{T1} is 10 percent, with an output signal calculated as follows:

T_{T1} optical power budget (with spliced couplers)

Transmitter output	−13	dBm
Connector losses	2	dB
Excess loss	0.2	dB
Tap ratio, 10%	10	dB
Waveguide loss	0.2 dB	
Power input to up converter receiver	−25.4	dBm

If each tap were equipped with fiber-optic connectors, then any signals passing through 10 or 12 taps would be attenuated below the minimum receiver sensitivity. However, suppose each tap were spliced in; then the loss would be calculated as follows:

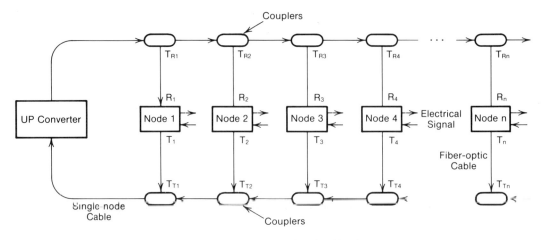

Figure 7–21 Fiber-optic version of a broadband cable plant.

T_{T1} Optical Power Budget

Transmitter output	−13 dBm
Splice loss out of coupler	−0.1 dB
Excess loss	0.2 dB
Tap ratio 10%	10 dB
Waveguide loss (1 km)	−0.2 dB
Connector loss into coupler	1 dB
Power input to up converter receiver	−23.4 dBm

In the preceding example, the coupler input is equipped with a connector, but the input from T_{T2} and output to the up converter are spliced into the bus, and each tap (coupler) represents a 0.4-dB loss.

Next, let us calculate the maximum loss between any transmitter and the up converter receiver, which is the difference between the transmitter output power and the receiver maximum sensitivity, or 29 dB. Note that these calculations are not worst-case because some of the devices are experimental. For the node to function correctly, the node must meet the criteria in the following equation:

$$5 \text{ dB} < \text{attenuation} \leq 29 \text{ dB}$$

The 5-dB minimum attenuation is required because the receiver dynamic range is only 24 dB. A further assumption is made that the dynamic range does not extend beyond the maximum sensitivity of the receiver.

Let us now examine the nth coupler in the system. Assume a 50 percent coupler is used; then the coupler output will be −16.4 dBm and the maximum loss from T_{TN} output to the up converter receiver must be less than 15.6 dB. This figure allows, for example, 63 taps to be installed on a 2-km bus without repeaters. If more than 63 taps are necessary, the same technique for calculating the receiver output power for the up converter may be used to calculate the repeater receiver input power levels. Repeaters are not shown in Figure 7–21 because they are similar to the up converter, with the exception that they do not translate the frequency up.

The up converter translates all incoming signals to the receiver band. For example, if T_{T1} through T_{TN} transmit a carrier in the 70- to 100-MHz band, which is FSK modulated, and the nodes use the carrier on a time-division multiplexed basis, the up converter translates the carrier to the receive band (226 to 266 MHz). The repeater reinforces the signal and is transparent to the nodes on the network.

At the receiver input of node 1, the input at R_1 from T_{R1} must be low enough to prevent saturation. Also, the tap ratios must be extremely large because a part of the signal must enter each tap and be monitored by the receivers. Therefore, each tap induces a loss of signal in addition to splice and excess loss.

Let us assume a tap ratio of 0.5 percent; then the first receiver's optical power input would be as follows:

Up convertor transmitter output	-13 dBm
Waveguide loss, 1 km	0.2 dB
Coupler splice loss	0.1 dB
Excess loss	0.2 dB
Connector loss	1.0 dB
Tap ratios	23.0 dB
Power input at node 1, R_1	-37.5 dB

If the assumption is made that 63 nodes are connected to the bus, then the loss without any power tapped from the bus would be 29 dB, and the nodes at the far end would be inoperative. There are two solutions; either add a repeater or provide a higher power source (laser).

The couplers used on the transmit side can be programmable (i.e., the tap ratios can be set using voltages to program the channels through the couplers). This type of voltage control applies only to single-mode technology. Fixed tap ratios would be difficult to adjust if a node and length of cable were added to an existing cable plant, for example, between nodes 1 and 2.

Overview of Packet Radio and Satellite

Packet radio and satellite communications will be necessary for local-area networks in hostile environments and where rights-of-way are difficult to obtain. A single example of a radio link is exhibited in large cities where two or three buildings are part of the same company. Radio links connecting the buildings over short distances could be used. Obviously, to lay cable through city streets would be very expensive.

Several large companies, such as General Motors, IBM, and Grumman, have private communication networks that are the size of small telephone companies. Satellites are used to connect company facilities between cities, states, and countries.

Hostile environments such as the polar regions make the laying of cable nearly impossible; therefore, satellite communications would be a necessity. Radio (terrestrial) would require repeaters, which would be exposed to the elements and be a source of link failure.

Radio and satellite both are vulnerable to atmospheric conditions. Typical voice and data traffic have BERs of 10^{-3} and 10^{-7}, respectively. For a BER of 10^{-3} in PCM voice transmission, the errors are perceptible. To improve the quality of radio links, frequency diversity and/or space diversity may be employed. Frequency diversity occurs when extra transmitters are employed that operate on a stand-by basis. If atmospheric conditions or transmitter failure occur, the secondary unit is automatically switched on. Space diversity involves adding additional receivers (spare paths).

If extra-strong signals are provided at the receiver under normal operation, radio transmission will be protected from fading, which induces errors. The higher the frequency of transmission is, the larger the required fade margin (extra signal strength). Normal microwave requires about 35 to 40 dB of fade margin, while 50 dB is required in the 10- to 20-GHz region.

Fading is dependent on atmospheric conditions that cause multipath distortion (i.e., when sky waves and ground waves are sufficiently out of phase to cancel). Heavy rainfall, fog, lightning, and humidity all affect fading.

Satellites have a large free space loss of about 200 dB, which must be overcome by the transmitter power and receiver gain. Communication satellites are located 22,300 miles above the earth in a geostationary orbit. The power output of the transmitter is approximately 5 W due to the expense of launch per watt. Thermal noise accounts for the majority of down-link noise, and intermodulation noise is the next contributor. The signal received at the ground station is -150 dBm, which of course is an extremely low level signal.

Associated with satellites is round-trip delay, calculated as follows:

$$\frac{\tau}{l} = \frac{2n}{c} = \frac{2 \times 1.0}{1186 \text{ ft/s}}$$

where $n = 1.0$ for free space. Then

$$\tau \cong 0.238 \text{ s}, \quad \text{for a round trip}$$

The transmitting station can check the data for collisions during packet transmission for digital systems. However, the receiver will not know until approximately ¼ s after it has occurred.

Sun-transit outages occur when a satellite is directly between the sun and the earth; a backup satellite must be provided if the outage presents a problem. This outage occurs for approximately ½ hour per day for several days around the equinox.

Both radio and satellite have applications that are not conducive to fiber optics. Also, certain other applications are well suited to fiber optics and cannot be implemented with satellite or radio links. But the networking technologies presented have cost and performance trade-offs when more than one technology can be implemented.

Satellite and radio links have been in existence for some time; therefore, numerous papers and articles can be found on the subjects. The design aspects will not be addressed here as they are beyond the scope of the text.

Selection of Local-area Networks

A key issue when selecting a viable network is whether it will transfer the data at an adequate rate; the maximum number of network nodes must be determined and the maximum distance between nodes. No single LAN meets all the criteria, but one of them may provide the best performance.

CSMA/CD networks perform the best where data transmission is bursty and frame arrival is not critical. As collisions increase during heavily loaded conditions, the performance of these networks begins to degrade. For example, after a frame is sent with no collision, it may collide with other frames as it progresses along the network, which can cause many retransmissions before the frame reaches its final destination. Channel capacity will drop to low levels as this overhead increases (approximately 30 percent). Therefore, the peak and average loading on the network must be considered.

If peak loads are large and cause undue delays on the network, token-passing bus or rings may become more attractive. Prioritized frame structures may allow only data with high priority to be sent during peak load periods. All frames may be sent at equal priority during low traffic load periods. This, of course, adds complexity to the nework (i.e., traffic must be monitored and a load assessment made).

Another factor that must be considered is the maximum length of the network. This parameter affects frame size, token hold time, CSMA/CD transmission timeout, packet service time, and so on.

The medium is also a factor that affects network length. Twisted pair, coaxial cable, fiber-optic cable and air each have propagation delays, as follows:

Medium	ns/m
Optical waveguide	4.97
Coaxial cable	4.33
Air terrestrial link	6.0
Satellite	3.3
Twisted pair	5.13

The coaxial and twisted pair are Ethernet cables, which is fairly representative of the two types of cables. Satellite propagation delay is assumed to be transmission through a vacuum; therefore, the delay is a little larger. Coaxial and twisted pair cable are dependent on the characteristic impedance parameters.

These parameters are useful for a first design iteration. In this section some of the physical layer parameters were discussed as well as networks. Often, the material must be covered together to assure the continuity of the subject matter.

PROBLEMS

7–1. Complete the routing table for nodes 3 and 4 in Figure 7–2(a).

7–2. Draw a routing graph for nodes 3 and 4 in Figure 7–2(a).

7–3. Configure an Ethernet network similar to a ring and spoke, where the center or hub is a series of 5 VAX 11/780 with 10 terminals and spokes are links interconnecting three other buildings besides the hub. The remote nodes are all interconnected to form a ring. Draw a block diagram of the system.

7–4. Evaluate Equation 7–12 for the following values of the variables: $P_0 = 10$ μW, $r = 0.5$ A/W, $G = 10$, $m = 50\%$, $B_n = 300$ MHz, $\alpha = 0.5$, $F_T = 0.5$, $\lambda = 820$ nm, $I_d = 10$ nA, $I_L = 5$ nA, $g_m = 2000$ μmhos, $C = 1$ pF. The star is an eight-port device.

7–5. For the previous problem, is there an advantage to using a 16-port device?

REFERENCES

[1] Beardsley and others, "On Survivable Rings," *Proceedings of FOCLAN*, 1982, Gatekeeper, Inc.

[2] J. R. Pierce, "Network for Block Switching of Data," *Bell System Tech. J.*, July–August 1972,

[3] Jerome Saltzer and K. Progren, "A Star Shaped Ring Network with High Maintainability," *Computer Networks*, vol. 4, 1980, p. 246.

[4] H. Salwen, "In Praise of Ring Architecture Local Area Networks," *Computer Design*, March 1983,

[5] B. Stuch, "Calculating the Maximum Mean Data Rate in Local Area Networks," *IEEE Computer*, May 1983,

[6] Technical Staff of CSELT, *Optical Fibre Communication*, McGraw-Hill Book Co., New York, 1980, p. 739.

8
TRANSPORT LAYER

The transport layer completes the data communication functions (i.e., it is the last layer related to hardware or networking). The next three layers are generally data-processing layers.

At the network layer, each machine is a single entity, but each machine may have multiple processes communicating with each other. The transport layer provides the necessary multiplexing.

Transport facilities provide two functions:

1. Interprocessing of communications between local process and remote node processes.
2. An additional error-checking capability.

Figure 8–1 is an example of a transport layer connection. Note that the network connection is a single-path copper or optical fiber, and multiple processes are serviced with required communication channels. Local communication at node 1 requires terminals 1 and 2 and the printer to communicate without using the ring. For remote processes to communicate, the ring is used (i.e., terminals 2 and 3, 1 and 4, 1 and 3, etc., can be communicating on a time-shared basis due to the transport layer).

The transport packets are embedded in the network packets in the data segment. These transport packets require a header with information similar to network packets. When the processes are communicating, the network packets are transparent to them (i.e., each process transmits and receives data as though it were the only network user). The data rate will be significantly lower when multiple processes are communicating due to the time-division multiplexing of the transport service. The transport layer produces virtual data channel service to a single network connection.

Transport Layer

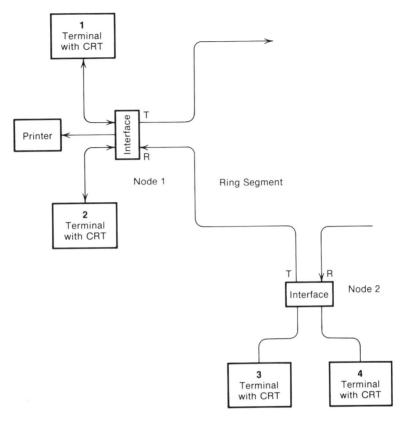

Figure 8–1 Typical LAN that requires a transport service. The node is considered the transport station.

When a single process is required for a node, such as a single terminal connected to a network, it is referred to as a connection machine. Then, obviously, no transport service is required because only a single process is connected to the network.

The transport addresses are often referred to as sockets. Addresses that require transport service will have an address for the network and one for the process or port. One problem with this technique is that the address field for source and destination is significantly increased, which adds overhead bits to packets and, of course, reduces throughput.

ARPANET TCP protocol, for example, provides 32-bit fields for network and host with 16 additional bits for ports. The CCITTs addressing scheme provides 4 decimal number fields for the network number and 10 digit fields for the host and port. These large address fields in large systems can be reduced by using a short-connection ID. This is an ID assigned after a virtual connection is established. This technique requires connection tables in both the host and network that have a directory ID number to corresponding true addresses. The shorter address and header length is a trade-off for increases in storage and the processing capacity of the node to generate connection tables.

The well-known port technique is used to provide the user with a particular service. The method of implementation is as follows: Directories with the service are published with their appropriate port numbers. Any user desiring these services may address the particular well-known port. For example the ARPANET TCP well-known ports accept these service requests directly and NCP well-known ports can participate in only one connection at a time, so requests must be passed off to another unused port. The well-known port is again ready to accept another request. This technique is simlar to making an initial connection on one port and transferring to another, which will allow the first port to be used again for an initial contact.

DECNET and SNA provide a directory so that the service may be called out by name. The directory locates the service, if it exists, and then makes the port connection or assigns a port to the service. When the service is not available, the request is rejected. SNA has a centralized directory service, while DECNET has a distributed version. The latter service is distributed among network nodes.

Sytek (a broadband cable system) has a two-port addressing scheme. The first four digits address the PCU, and the port within the devices is identified by the last digit. Each FDM channel supports 200 active users, and the system has 120 channels in groups of 20; therefore, the total number of simultaneous active users is 24,000. Sytek has a rotary channel selection technique that allows the user to dial an address within the group. A connection is made with the first available port within the group. This technique allows a block of addresses to be assigned to a network implement such as a host computer, which is much more efficient than assigning a single port.

Packets are the fundamental unit of data transfer, with the additional transport header. The transport protocol must break up large messages into packet-sized segments. See Figure 8–2, a typical transport packet. The header must contain control information such as whether the packet is a system or process packet. The transport station may send out packets for flow control or convey urgent data for time critical data transmission. It usually carries only one byte of data. The control field may also contain information as to whether an ACK is required, an attenuation bit, and an end of message (EOM) bit. Urgent data packets have the attention bit set, which alerts the processes to prepare to accept urgent data (receiving station).

A data-stream type of field may be provided for session layers or higher-level protocols. This field is not examined by the transport protocol. Source and destination process identifications are necessary in the header also. These are self-explanatory. Sequence numbers are required to assist the transport protocol in assembling messages after the packets are received. Acknowledgment numbers are also required to inform the transmitting station as to which packets have been successfully received. This number may be sent for individual packets or groups of packets.

Transport Layer

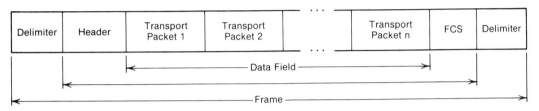

Figure 8-2 Typical frame with transport packets.

Some systems allow the transport stations to allocate their own connection identification. The caller is notified in the first replay packet. This interchange of information may be carried out by the transport stations to form a virtual connection with the use of system packets. Both the originating and receiving stations assign ID numbers under system control. The advantage of this technique is that buffer space can be made smaller, directory service is dynamic (i.e., the directory includes only processes that are in communication), and processing time is reduced because the directory of ID numbers is smaller. On the negative side, a more complex algorithm is required to accomplish the dynamic directory services.

Transport Error Recovery

A rather formidable problem in most communication is error recovery. If packets are damaged or lost, the communicating stations must have the capability to negotiate the retransmission of the data. Initially, however, the receiving station must be aware that an error has occurred or a packet is lost.

Error detection is accomplished in the following manner: The transmitting station makes a calculation on the data, which results in a checksum; the checksum is included in the packet. When the packet arrives at the receiving station, the identical calculation made at the transmitter is made at the receiver, and the checksums must match. If the transmitted and receiver calculated checksums do not agree, an error has occurred. Some systems, such as Ethernet, have large checksums of 32 bits. Burst errors must be greater than 36 bits before they will be undetected. The bit error rate is somewhere in the neighborhood of 10^{-10} to 10^{-12}. However, this represents additional overhead.

When packets are received by the transport station undamaged, an ACK must be sent to the originating station, and this frame must also have a checksum. The situation just described implies that the ACK may have been acceptable to the station, but it may or may not be acceptable to the process, which implies damage between station and the process.

When packets are received with errors, a NACK must be transmitted by the receiving station. However, a problem exists because ACKs and NACKs may get lost or damaged in the subsystem levels or a considerable length of time may lapse before they are received at the originating station. A method to alleviate this problem is to send a frame and start a timer, and if an ACK is not received before the timeout, the data are retransmitted. The NACK can also be used in conjunction with this technique to stimulate retransmission. A method is also necessary to stop retransmission should no ACK be received after a number of tries. The connection should be closed down after the users are notified.

The timeout for retransmission is a critical parameter that affects performance. For short timeout intervals, larger numbers of retransmissions occur, and this reduces channel capacity. If the interval is too long, the mean transmission delay increases if there is a significant error rate. For optimum performance, the timeout interval should be just long enough to receive an acknowledgment under normal operation. But delay in packet-switching networks can vary widely. Some trade-offs can be made by using the efficiency equations of Chapter 5.

Retransmission of packets can cause some problems at the receiving end if packet identifiers are not used. Packets arriving at the receiver are checked against the identifier list for duplication of packets. The duplicate packets are discarded. The identifier field must be large enough so that when the values are used over again different packets do not have the same identifier. The identifier field size is dependent on the maximum packet lifetime (i.e., the maximum time a packet may exist in the transmission medium).

Sequence numbers, when used to identify packets, may occur in a block form so that the receiver need store only a single number. This single number indicates a packet and all its predecessors. For the situation when packets are delivered out of sequence, a small number of individual sequence numbers can be dedicated for this task.

Flow control and buffering limit the rate of data transfer. At the transport level, the receiver most often limits the originating station activity. If the receiver can process data at a fixed rate, no explicit flow control is necessary. Simple flow control can be achieved with a receiver that issues start and stop commands to the sending end. However, this type of flow control operates poorly when transmission delays are variable. For example, the stop command may be followed by data and start commmands followed by large delays before new data arrive. The receiver will be confused as to when start and stop commands have taken effect.

A more effective flow-control technique can be initiated if the transmitter and receiver agree on the amount of data to be transmitted. The receiver grants "credits" to the transmitter; when they are used up, transmission stops. The negotiations for credits are conducted through the use of special control packets. The method described is absolute; that is, when a packet is lost, the transmitter may stop transmitting (end of credit) while the receiver is waiting (lost credit). To prevent this problem, both credits and sequence numbers must be negotiated by control frames. Credits can be made relative to sequence numbers; this technique improves error tolerance over the previous method.

A more common term, a window, relates to the number of credits a transmitter receives. This technique has a large advantage: if window is specified by a block of sequence numbers, it can be adjusted in the following manner rather easily. As the number of acknowledgments at the lower end of the window is received, the window can be reduced in size. Also, the window may be increased in size by simply increasing the sequence number value. Thus window size may be adjusted by fixing an upper or lower sequence number and adjusting the other.

Buffer space is often closely related to window size. If window size is too large, buffer space may become filled and packets will be discarded. Thus large numbers of retransmissions will be necessary, which decreases throughput. Small window size constrains the sender in some cases from using the peak transmission rate; therefore, throughput again suffers.

The previous discussion describes a method of flow control between the transmitting and receiving functions of the transport protocol. This is not, however, congestion control, which is the mechanism the network invokes to protect itself from excessive traffic. Thus the transport station may also limit the flow over one connection to allow a fair allocation of facilities to all the transport processes.

Some limit on buffer space is necessary for economic and perhaps physical reasons. Large buffer space may be required due to the need to store out-of-sequence packets and for matching receiver and transmitter transmission rates. When packets arrive out of order or packets within a sequence are lost, the buffer must store the packets until the predecessors have arrived, rather than discard them and require retransmission from the sender. Mismatched transmission rates, variable transmission delays, and differing processing capabilities all lead to uneven arrival and receiver acceptance times; this requires some sort of buffering to prevent blocking. Thus flow control is closely tied to buffer availability. Transport protocol performance is therefore very dependent on buffer space availability, also.

Synchronization is often required for real-time processes. Sequence numbers do not reflect any timing requirements for packets, only the order of transmission. Graphics, speech, teleconferencing, and process control all require real-time packetized transmission if local-area networks are to be implemented. Each packet requires some type of time reference or time stamp to maintain correct synchronization.

Real-time LANs have some unique problems. For example, the transmitter and receiver of the originating and receiving stations must have clocks that run at the same rates. The receiver has a delayed version of the transmit data. The delay prevents the newly arrived data from being immediately delivered to the process; this is a form of time buffering.

In some real-time applications, timing of data is more important than completeness. For example, packetized speech transmission requires that packets with long delays be simply dropped. Normal ACK, NACK, and retransmission mechanisms may also be dropped. Occasional missing packets represent only a single data sample for speech, and the error may be nonprecipitable.

Teleconferencing makes some unique demands on a transport facility. For example, text graphics and speech require synchronization and possible transmission over parallel data streams. On the other hand, when half-duplex broadcast situations occur, large delays may be possible (data can be recorded on disk and reproduced later).

For the situation where data and speech are using the same transport facilities, a method of prioritizing speech is necessary. Delays in data delivery do not cause any problems, but delays in speech cause synchronism difficulties, which result in poor reproduction at the receiver and associated process.

Some transport facilities provide an out-of-band channel that is not constrained by the normal flow-control mechanisms. This may be an interrupt-type control packet that has a small amount of data. For example, the X.25 protocol has 8 bits. Interrupt information can also be sent in a normal packet if one is to be transmitted. The receiving transport facility processes the data as they arrive with the interrupt. The interrupt or out-of-band request is passed to the appropriate receiving process to be executed. Out-of-band signals are used for interrogating receivers, performing housekeeping routines such as purging data from the receiver, or scanning the data. Out-of-band signaling can be similar to a maintenance order wire.

Network Security

Another function of the transport protocol is security. Normal protection provided by checksum and various network error-checking procedures is only useful under normal behavior. Some networks, such as military installations, banking institutions, and other networks that carry secret or industrial confidential data may require not only precautions for laying cable, but may also require data to be encrypted. Fiber-optic cable is very difficult to tap without detection. For industrial installations where national security is not at stake, fiber-optic cables with monitors and alarm circuits would suffice and no encryption would be required.

Encryption can also be added to reduce the risk of intrusion for common copper lines, whether they are coaxial cable or twisted pair. Several manufacturers produce integrated-circuit encryption circuits. The number of possible keys for these devices is astronomical.

The originating and receiving must both be issued with a common key to communicate. Also, keys must only be distributed to authentic network users (i.e., the distribution of keys must be secure). If an unauthorized network user has a key, all keys must be changed and reissued, which can be a monumental task for large networks. Only encrypt what is necessary (i.e., control information, addresses, and sequence numbers may not need encryption). The techniques for making data secure in hostile environments (during wars) are beyond the scope of this book.

Transport-layer Efficiency

When host computer, terminal modems, and the like, exhibit a failure, the network needs initialization to establish orderly communication. Transport-layer facilities provide this service with special control packets. System computer failures are one of the most serious types, especially when soft failures occur such as RAM voltage drops, which cause erroneous data in buffers and main computer memories. These types of failures can go undetected unless some form of memory test is made.

Transport-layer efficiency is rather difficult to analyze if all processes do not use the network channel equally. For example, one process may transfer files and another may be a low-data-rate transfer to a terminal. All the processes may not have equal access; they may be prioritized or statistically they may be different.

Efficiency is dependent on overhead, which results from checksums, addresses, sequence numbers, packet length, window size, ACKs, NACKs, and the like. The reliability of the channels is increased at the expense of throughput. Retransmissions add to throughput reductions and delay because data are transmitted more than once in their entirety. Flow control also forces the channel into idle, which adds further inefficiency to commnication between processes.

Many items that compose the header of a packet are fixed in length. As packet length increases, throughput increases, but delay also increases because the processing time for each node increases. Efficiency increases also because the percentage of the packet that is the header becomes smaller. The reverse is also true; smaller packet size results in lower efficiency, less delay, and lower throughput. Efficiency has been covered previously except for the transport processes, which can be treated for a first-cut analysis similar to the network case.

Retransmission in transport protocols has a large impact on protocol efficiency; it is necessary when packets are damaged or lost. The damage may occur in the medium or during the transfer from transport station to process. A retransmission interval when made large does not waste bandwidth and increases throughput. While for the other extreme (i.e., small retransmission intervals), delay is reduced because damaged packets are retransmitted sooner; therefore, bandwidth and throughput are traded for smaller delay.

For the sequencing of packets, as well as error correction, the assumption that delays are identical cannot be made. The loss of a packet or its ACK delays delivery of subsequent packets until the error is corrected. If a message is made up of several packets, the transport protocol may not deliver it until it is completely assembled. A NACK tends to reduce the problem somewhat because it may be acted on immediately with a retransmission of the damaged packet. The adjustment of various parameters may be necessary to optimize performance. For further information and calculations of transport protocols, see reference [1].

Transport-layer Hardware Issues

In Chapter 7, a DEC network terminal server was discussed. This particular hardware unit has a transport protocol that time division multiplexes up to 32 terminals, with transmission rates of 50 bits/s to 19.2 Kbits/s. The terminals can operate in the full-duplex mode. This terminal server communicates with Ethernet hosts that implement local-area transport (LAT) protocol. For standard networks such as Ethernet and Wangnet, implements may be obtained that have the transport layer built into the device. Some other transport features are examined in the following paragraphs.

The server software is loaded from a host into the server memory; this includes diagnostics for maintenance purposes. If a server malfunction occurs, the server dumps the memory image and automatically reloads the software. The memory image dump is analyzed later by the diagnostics to determine the cause of the malfunction.

This particular unit is provided with a measure of security. It does not allow access to network hosts or other resources without an appropriate password.

Virtual terminal support is available on the network server. The terminals are connected by dedicated means or dial-up lines. The network server provides several services such as connection management, terminal parameter configuration (speed line width, fonts, etc.), and management facilities. The server also has RS-232/CCITT V.28 interfaces. This server supports a series of DEC terminals, such as VT100, VT200, and LA100 types. This device has a large number of software options. For further information, see the Digital Equipment Network Communications catalog.

Virtual Terminals

The virtual terminal will be discussed in Chapters 10 and 11 in more detail, but an overview is presented here. More than 100 different terminals are manufactured, and each may have peculiarities that make its repertoire of commands and transmit/receive facilities somewhat different from the others. Many manufacturers design their terminals to emulate a particular popular model. With the advent of personal computers, it is possible to emulate a multitude of various terminals. The virtual terminal protocol is devised to hide the differences between various terminals and make them all behave according to three broad classes.

The first is the scroll-mode terminal, which lacks any onboard intelligence (microprocessors). This class of terminal cannot perform any editing functions. When keys on the terminal are depressed, the data are sent over the network. As data are received by the terminal, they are displayed, and when a line of characters is filled, it is scrolled upward. These types of terminals are also often referred to as "dumb" terminals because they do not have a microprocessor.

The second class of virtual terminal is the page-mode terminal. This type has some editing capability. These terminals display a full CRT page of text (i.e., 24 lines by 80 characters of text). They are equipped with a cursor, which can be manipulated by the operator or the computer.

The third class of virtual terminal is the data-entry type. These devices are microcomputer based (i.e., they are equipped to do local processing). They generally have prompting and displays that are forms. Terminals of this type are used in sales offices, airline reservations, factory inventory, and other environments where sophisticated displays of data are required. The transport protocol must transfer single characters for the first class and files for the second and third classes. The protocol of the transport station may not transfer a character at a time but assemble lines or even pages into packets. Sending a character at a time is very inefficient.

Network Servers

The DECNET Router Server is an example of a device with transport protocol capability. This server connects via an Ethernet transceiver and transceiver cable to the coaxial Ethernet cable. The router server will connect up to eight RS 232C/CCITT V.28 interfaces that operate full duplex at 19.2 Kbits/s to the Ethernet cable, or eight 56 Kbits/s to CCITT X.35. This unit also supports 500-Kbit/s or two 250-Kbit/s synchronous CCITT V.35 line interfaces. These latter two features are router server options. Network router servers can be used to connect remote Ethernet cables, which allows the designer to construct more complex networks. The router server has an adaptive routing and network management algorithm that may be stored locally in the device. This relieves the host computer of this burden.

As a final example of a device equipped with transport protocol facilities, consider the DECNET Router/X.25 gateway. This router is similar to the previously mentioned equipment except it allows X.25 packet-switched networks to communicate with nodes located on Ethernet cable whether they are terminals or host computers.

ARPANET

A further examination of some common network architectures is warranted. One of the most important is ARPANET and its transport protocols NCP (network control protocol) and TCP (transmission control protocol). The NCP protocol is used in an error-free environment, and for full-duplex operation two physical paths are required. The first 8 bits of the message represent a control function, as follows:

NOP   No Operation
RTS   Receiver to Sender. This is a call request.
STR   Sender to Receiver. This is similar to call accepted.
CLS   Closed. This is a rejection of an attempt to make a connection.
ALL   Allocation. This is the message required at the sending host, which must be sent by the receiving host.
GVB   Give Back. When the receiver becomes swamped and wishes some of the buffer allocation promised to the sender, it sends a give-back control word (i.e., provided the receiver has sent an ALL control word).
RET   Return. The sender replies to a GVB receiver control word with RET and the amount given back.
INR   Interrupt by the Receiver.
INS   Interrupt by the Sender.
ECO   Echo. This message is used by a host to test if another host is running.
ERP   Echo Reply.
ERR   Error. This is an error message control word for assisting in troubleshooting software.
RST   Restart.
RRP   Restart Reply. These last two control words purge all links.

Let us now examine the NCP messages to observe their composition. The NOP, RST, and RRP are single 8-bit messages that are self-explanatory. The RTS and STR message format is shown in Figure 8–3.

Receive ports are even 32-bit numbers and odd numbers are send ports. Each connection is equipped with a send and receive port. The link number is sent with the RTS message to allow identification of the connection, which replaces the two 32-bit receive and send port numbers after the connection is established. This latter technique obviously reduces header overhead because only the connection number need be sent. The link number is also passed down to layer 3.

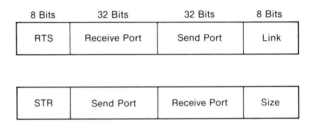

Figure 8–3 RTS and STR message formats.

Transport Layer

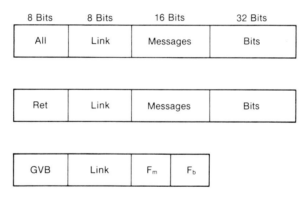

Figure 8–4 Close message.

The size of the sender's message must be known by the receiver to provide buffer space.

The next control message to consider is the close message shown in Figure 8–4. The socket information is a session-layer entity that is passed up to layer 5.

The next three control messages to be considered are ALL, GVB, and RET, as shown in Figure 8–5. After the handshaking results in a connection, the ALL message is sent to the sending host on link zero. The buffer allocation is on the host-pair basis, with the message and bit limit sent in the appropriate field as shown in the diagram.

Figure 8–5 ALL, GVB, and RET control messages.

The GVB requests the sender to give back some of its allocation (ALL) with a fraction of the message and bit allocation. The sender responds with a RET message.

The INR and INS are both messages with the link number included (i.e., they are 16 bits long).

Echo (ECO) and echo reply (ERP) both require a character field (16-bit total length, control word plus character).

The last control message format is the error message. Not only will the error be indicated but the code is also sent to identify it. A data field is provided to comment on the error details (e.g., bad parameters or ASCII code error).

For ARPANET with TCP, the network is considered to be lossy. TCP transport stations are made to communicate over networks with low reliability, (i.e., messages may not always be delivered intact). The packets are constructed with

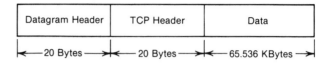

Figure 8–6 Datagram and TCP.

the network layer delivering data (i.e., it gives no guarantee that datagrams are delivered promptly or in order). The TCP protocol arranges for retransmission and assembles the message after all segments have arrived correctly. The datagram and TCP are shown in Figure 8–6.

Table 8–1 is a description of the datagram header components, and the header is shown in Figure 8–7 with symbol designations.

TABLE 8–1
Datagram Header Fields

Symbol	Field Name	Field (bits)	Comment
V	Version	4	Allows the use of multiple protocols.
HL	Header length	8	Specifies header length.
T	Total length	16	Header + data = total length.
NO.	Datagram number	16	Destination host identifies which datagram a fragment belongs to.
DF	Do not fragment	1	Instructs gateways not to fragment data because destination station cannot reassemble it.
MF	More fragments	1	Message is fragmented; needs more to assemble it.
F_{off}	Fragment offset	13	All fragments are a multiple of 8 bytes except the last. This word describes which fragment is represented.
T.O.	Time out	8	This is the packet lifetime timeout (255 s).
P	Protocol	8	This field indicates which transport station the datagram belongs to. The datagram header has no port numbers.
CS	Checksum	16	This checksum verifies the datagram header only.
SA	Source address	32	First 8 bits indicate network number, and next 24 bits indicate host number.
DA	Destination address	32	Same as source address format.
OF	Option field	32	This field can be 0 or more 32-bit words. This field is used for security, error reporting, maintenance, time stamping, and other miscellaneous information not provided for in the header.

Figure 8–7 Datagram header with symbol designations.

Figure 8–8 TCP header format.

Next consider the TCP header with the descriptions given in Table 8–2.

The TCP header format is shown in Figure 8–8. As one can observe, TCP protocols allow for message fragments that are out of sequence and they have an ACK field. The data are prepared at the transport station and the header, and the TCP header is appended to the data. The datagram header is attached to the TCP header prior to transmission over the network.

TABLE 8–2
Transmission Control Protocol TCP Header

Symbol	Field Name	Field (bits)	Comment
SP	Source port	16	Identifies processes.
DP	Destination port	16	Communicating.
SQ	Sequence number	32	Usual transport sequence numbering technique.
ACF	Acknowledge	32	Acknowledges by piggybacking.
HLT	TCP header length	4	Self-explanatory.
URG	Urgent pointer	1	Urgent pointer in use.
ACK	Acknowledge field	1	ACK = 0, not in use; ACK = 1, field in use.
EOM	End of message	1	Self-explanatory.
RST	Reset	1	Reset connection due to confusion, host program inoperative, large errors.
SYN	Synchronism	1	Synchronization bit is used to establish a connection; may be considered also a connection-request bit.
FIN	Finish	1	Used to close down the connection (i.e., sender has no more data).
W	Window	16	Gives the number of bytes beyond the ACK byte.
CST	Checksum TCP	16	Checksum of data and TCP header.
UP	Urgent pointer	16	Indicates by a byte offset from the current sequence number which urgent data are present.
OP	Option field	32	Zero or more 32-bit words. This field is used for miscellaneous purposes. The field may be used for security, buffer size, coding/decoding, etc.

DECNET

The equivalent to the transport protocol is the network services protocol (NSP) for DECNET. Nodes are all treated alike, similar to the ISO layer, but ARPANET does have a distinction between host and switching nodes.

Addressing in DECNET may be to specific ports or done indirectly by using names and an address directory that is a function of the routing table. This applies to both source and destination addressing. Protocol compatability is ensured through an exchange of protocol version numbers. Flow control and error control are accomplished on a message basis with agreement between sending and receiving processes. The credit agreement between the sender and receiver is a conservative one (i.e., the receiver only grants credits for the buffer space available). Messages may be broken up into segments by the source, and no further fragmentation may occur during transmission through intermediate nodes.

CCITT X.25 Packet-switching Network

This network standard specifies the interface between subscriber computer (data terminal equipment) DTE and the local network attachment (data communication equipment) DCE. The physical link is managed under a version of HDLC protocol and an upper layer virtual circuit protocol. This transport protocol is similar to layer 3, rather than layer 4, of the ISO hierarchy.

This protocol performs some of the functions of layer 4, such as packet sequencing and flow control. Some out-of-band interrupt signaling is provided, which is a normal function of the transport station under ISO layering. Thus above the network layer, networks require further definition. X.75 can be considered when virtual circuit service is required. This standard is similar to X.25, but it specifies virtual circuit protocols for links between networks with virtual circuit service for internal calls.

For lack of space, all networks could not be examined. Prior to any design implementation, the reader should study some of the other networks, such as SNA, Wangnet, Sytek, and IEEE 802 standards.

REFERENCES

[1] Franklin F. Kuo, *Protocols and Techniques for Data Communications Networks*, Prentice-Hall, Inc., Englewood Cliffs, N.J., 1981, 35–77.

[2] A. S. Tanenbaum, *Computer Networks*, Prentice-Hall, Inc., Englewood Cliffs, N.J., 1981, 324–382.

9
SESSION LAYER

The session layer function provides the interface between the hardware and software in most cases. The transport layer must communicate through the session layer with the presentation layer, which is software. The presentation layer (discussed in Chapter 10) must negotiate through the session layer to establish a connection with a process on another machine. Session-layer addressing must map directly into the transport-layer addresses. The session layer performs the necessary handshaking to allow the connection to be established between two processes. This layer performs the necessary management of the session after it has been established. The session layer maintains itself intact if lower subnets fail or retransmission is required in the transport layer due to lost or damaged packets. The session provides the necessary smoothing of any hardware problems so that they are transparent to upper network layers. Of course, there are limitations; for example, if a connection is broken for any length of time, the session layer cannot maintain the necessary continuity. Not all networks have this layer, or it may not be well defined as presented here in the ISO model. The session layer is often embedded in the operating system as software drivers.

The session layer can best be described by examining some of the commonly used LANs. As will be noted in these examples, the network layer will not be as clearly defined as the layers above and below it. This is because this layer takes the bits provided by the hardware (transport) and makes a user-oriented service from it.

The session layer provides the conversion of the source address from remote users to an identification that the host can recognize. When the identity of the sender is known, the host operating system can deal with it accordingly (e.g., it may not be in the directory of host users or have only limited service privileges).

For situations when processes communicate remotely, log in is required from the user to the host process. Log in may be required several times, which would be unproductive for the human operator at the remote site. The session provides this service, which is transparent to the operator. This is not to be confused with log in by the human operator.

The session layer can also segment long file transfers. For example, if the transport layer receives a long message involving many pages of text, the session layer can break this up into segments and demand an acknowledgment after each. If a failure occurs in the network, the total file need not be retransmitted, only the segments that are not acknowledged. Also, if a transport failure occurs, the session layer will make it transparent to the host and remote user. This implies that the session layer can reestablish the transport connection automatically, which is usually the case.

The reader can examine what has been established thus far in the ISO model. First, a physical medium is determined with some physical characteristics; this medium must be made to connect and operate in a network, which is the function of the data-link layer. The physical point-to-point connections are then made to function as virtual connections, which expands network capability. This hardware must be made to function with the software; the session layer performs this task. As the upper layers are discussed further, one may observe their behavior—the operating system communicating with the program and finally the program communicating with the human operator. As the complexity of these various layers increases, the tendency will be to further divide the layers, which is actually the situation.

SNA Session

The SNA transport layer was not discussed in the previous chapter because it is difficult to separate from the session layer. The transmission control layer overlaps the transport and session layers of the ISO hierarchy (i.e., it performs some of the tasks of each layer). Also, note that the path control performs some functions that are a part of the ISO transport layer.

A simple diagram of an SNA network is shown in Figure 9–1. The logical units (LUs) in the diagram function as ports to network subscribers. The session layer communicates LU to LU. The SSCP is the central control for SNA network domains, as shown in the diagram. Connecting domains is similar to star network interconnections.

If a connection is desired between processes (LUs) (the process may be referred to by a symbolic name), the SSCP will map the name into a network address and arrange the session. The two domains that are communicating are unequal in their ability to communicate. One is designated the primary (it is the master), and the other is the secondary (more like a slave).

Within domains are a variety of nodes, resources, and also backup SSCPs. Sessions may be set up between LUs within a domain or between two LUs within

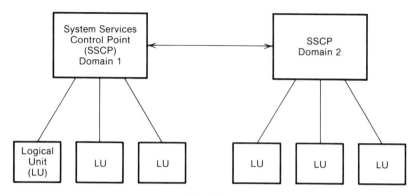

Figure 9-1 SNA network representation with all the common blocks shown.

different domains. Several domains may be included in larger networks. One reason for the complexity of SNA networks is due to the early implementation of single host computers. Many implements were added over a period of time until the complexity of these LANs had to be dealt with on an evolutionary basis.

Logic units are similar to virtual terminals (i.e., they have architectural similarities and are classified by type). The types are the following:

Type	Description
0	This type is defined by customers or for products.
1	Represents a keyboard–printer console with optional features such as printers, diskette drives, and other similar implementation.
2	The LU has the appearance of a keyboard display with local copy option.
3	This unit has receive only (RO); it is a printer or similar implement.
4	Similar to type 1 with a different choice of data stream.
5	Undefined.
6	This is probably one of the most important types; it is used for general-purpose transactions and supports distributed (array) processing. These units may be used to invoke remote access to queues, data bases, time-scheduled program execution, and other important attributes.
7	This type is similar to one with data format differences.

For further information on SNA LU types, see reference [1].

Let us now examine the sequence of events to establish a session. A total of 14 control messages is required to accomplish this task. The events are as follows:

Event	Description
1	The secondary LU communicates with the SSCP its desire to establish a session.

2   The SSCP secondary communicates with the primary SSCP that an LU wishes to establish a connection with one of its LUs. The secondary SSCP will provide the LU originating the session's password, to be authenticated by the primary SSCP.
3   If a remote SSCP has an authentic password and the primary is not saturated with sessions, the originating LU gets an ACK.
4   The secondary passes on the ACK to the originating LU.
5   The secondary SSCP sends a request to bind (activates a session between LUs) to the primary SSCP.
6   ACK is sent from the primary to the secondary SSCP.
7   Primary SSCP initiates binding at its LU.
8   ACK is sent from LU to the primary SSCP.
9   The bind command message is sent from the primary to the secondary LU and specifies the parameters for the session, such as protocol, flow control, and others that affect session performance.
10 The secondary ACKs the session data from the primary.
11 LU to primary SSCP starts the session.
12 Primary SSCP ACKs the command.
13 Primary SSCP to secondary SSCP session started.
14 ACK is sent from secondary SSCP to primary that the session has started.

The session setup procedure just described is a cross-domain type. Obviously, it is less difficult to set up a session within a single domain between two LUs using a single SSCP. This exercise will be left for the reader; see reference [1] for further information.

For termination of a session, the secondary LU can log off and the unbind is unconditional; it occurs immediately. If session termination is conditional, the secondary LU waits until the primary LU agrees to end the session. The method of termination is dependent on the types of LUs that are communicating.

SNA has a number of maintenance features that will be briefly described here:

1. Maintenance data are recorded as to the state of both the network and its components. These data are available on request by users.
2. All network operators receive failure notification when links or node failures affect their performance.
3. Damaged sessions are resynchronized through failure notification.
4. Provisions are made for backup and recovery.
5. Restart of a session after a failure begins with a half-session (half-duplex) resynchronization.

Session management and dialog control are initiated through the use of a 3-byte request response header as shown in Figure 9–2.

Figure 9-2 Request/response header.

Bits	
0	Request = 0, response = 1
1, 2	00 = pass message to network address unit 01 = network control message 10 = dialog control normal message 11 = session control message
3	Undefined bit
4	When set, indicates function header present in data field
5	When set, indicates the presence of a sense data field for error reporting
6, 7	00 = middle of a message chain (group of messages used for error recovery and other miscellaneous purposes) 01 = end of chain 10 = start of chain 11 = single message chain
8	For upper-layer ACK use
9	Undefined
10	For upper-layer ACK use
11	For upper-layer error use
12, 13, 14	Undefined
15	Flow field used by senders for permission to transmit another group of messages, and receivers use this field to grant or deny permission
16, 17	Limits bracket structure (groups of requests and responses in both directions constitute a bracket similar to chaining of messages)
18	Used to change direction when operating half-duplex
19	Undefined
20	Selects one of two character codes
21, 22, 23	Undefined

The previous discussion has given the reader an overview of the SNA session layer and some of its attributes.

Session Discussion

The session layer in both DECNET and ARPANET is conspicuously absent. DECNET's application layer deals directly with the network services layer (transport layer equivalent in ISO models) and ARPANET Telnet communicates directly with the host–host layer. These layers for DECNET and ARPANET will be addressed in Chapters 10 and 11, respectively.

Sytek Session Layer

The session layer for the Sytek broadband cable system will be investigated; it is quite similar to the ISO model. Addressing is easily accomplished using two component addresses; the first four characters identify the packet communication unit (PCU) and the next character identifies the port. This method of addressing the PCU is called fully qualified addressing. The second type of addressing is referred to as partial addressing, where only the first address in a group of unit addresses is given and a rotary call looks for the first free session in the group; this will be the port used by the session.

After addressing by either type previously mentioned, the PCU initiates a session. The session may be with a port within the same PCU or with a remote PCU. Remote PCUs can even establish sessions between two other PCUs. This allows a multiple remoting capability by a single node. A permanent session (e.g., between a host and terminal) can be established with a command that sets up the session automatically during periods of no PCU activity, or it can be triggered by an external event. When a call is made to a PCU, the sessions are numbered automatically, beginning with 1.

The PCUs are connected to the user equipment through an EIA RS232 interface. A succession of events must occur before a session can be established. An unprivileged session requested cannot be made to a remote if the data terminal ready (DTR) line is not asserted or is not asserted before a specified timeout. An override for the DTR line is the privilege attribute, set to ON within the PCU; it will ignore the DTR line. When a session is established (first session), the data set ready (DSR) line is asserted. For a situation when a PCU is active and no sessions are present, if an incoming session is presented the ring indicator (RI) drops to zero after 1 second.

Data carrier detect (DCD) is set to ON when a port has an active session. DTR, DSR, RI, and DCD are all pins on the EIA RS232 interface with directions of communication as shown in Table 9–1. The table indicates the direction of PCU and devices signaling (e.g., the PCU must handshake with a terminal prior to transmitting or receiving data).

Session termination is accomplished with a DONE command from the PCU or if the timeout between character transmission is exceeded; this value is specified for the local or remote PCU. If the DTR line is dropped low by the device connected to the PCU port, DSR drops low after the last session is terminated and is reasserted. Each PCU can handle multiple simultaneous sessions, which is 4 for LocalNet 20/100 and 16 for LocalNet 20/200. The sessions need not be allocated equally for the LocalNet 20/200 (i.e., port 1 may have a limit of 4 or 5 and port 3 may only have a limit of 1). These limits are set prior to any communication between PCUs. The total number of sessions cannot exceed the maximum number

TABLE 9–1
RS232 Pinouts

Pin	Name	Device	PCU	Function
1				Chassis ground
2	TD		→	Transmit data
3	RD	←		Receive data
4	RTS		→	Request to send
5	CTS	←		Clear to send
6	DSR	←		Data set ready
7	SG			Signal ground
8	DCD	←		Data carrier detect
20	DTR		→	Data terminal ready
22	RI	←		Ring indicator
25	AB		→	Autoband request

the PCU can support, and each port must not violate the maximum for that port.

Sessions may be suspended and switched (i.e., a device can request its PCU to switch from one session to another). The initial session remains active until explicitly suspended when the session is switched to a new session. After the switch has been made, the initial session can be suspended. If the session switched to had been inactive, it would have been reactivated.

When a device requests a new session while one is open, the completion of the new session request switches the device port to the new session. If a session is initiated to a remote port with a session in progress, the remote port is switched to the new session; but for the situation when the remote port is in data transfer mode, it does not switch to the new session. The latter provision prevents the loss or destruction of data during a transfer. This system also provides remote command execution and is a powerful tool for various maintenance, remote implement, and host access operations.

The discussion of the session layer as presented here will assist the reader in understanding some of the new implements that may interface with the session layer of the ISO model architecture.

REFERENCES

[1] Systems Network Architecture Formats and Protocol Reference Manual, Architecture Logic, IBM Systems Library, Order Number SC30–3112.

10
PRESENTATION LAYER

The presentation layer is an interface between the process and the application layer. This layer provides the file transfer, security, virtual terminal configuration, text compression, voice recognition, voice synthesis, code conversion (e.g., hexadecimal to ASCII), statistical traffic monitor, and other services that can be a part of the operating system and transparent to the application layer. The presentation layer may also contain library routines.

The chapter will encompass some of the presentation-layer attributes mentioned in detail, while others will be dealt with rather superficially. As an example, voice recognition is a subject with a high degree of complexity; this subject cannot be covered in any depth because of the large number of coding techniques used.

File Transfer Protocols

File transfer protocols define a set of rules to transfer files from one host file to another. The two hosts may share a disk storage station, which is a virtual file to each, or they may each have their own located on the LAN. Hosts may also be equipped with disk storage, which is part of their hardware.

Files may be transferred from one host to another, transformed, or manipulated. The first of these functions is the easiest to accomplish because the data are not altered but only moved from one host to another. The protocol allows the network subscriber to open a file (file name) and access the data within it. Data can then be transferred to and from the file and close the file. The operation described occurs under an access control mechanism. File transfers may be all or part of a file depending on the particular protocol.

For a file transfer from one machine to a storage medium, the data must be reproduced exactly upon retrieval. The source and destination machines may have differences in word length, character set, and the like.

File transformation, on the other hand, requires converting a file from one format to another. Remember, file transformation is not similar to language translation (e.g., English to German). During language translation, some of the information content of the source language is lost. In file transformation, the data from source to destination must be exactly the same.

The transformation may convert the source file into virtual form and the transfer will occur. This virtual file is then converted to a proper local file format at the destination. The virtual file technique allows all network implements to map their files into a virtual protocol that the network can deal with.

File manipulation is the objective after a file transfer has been made from storage to user, such as create, delete, rename, change password, or generally to make changes to the data.

File transfer protocols can be explained using the simple two-ring network in Figure 10–1. The disk storage resides on the corporate LAN ring; the host computer and terminals store data there. Each implement maps its file into a virtual file, and when it is received at the disk storage station the file is mapped from the virtual format into the local format. However, when data are sent from the satellite business office, the virtual file is transferred across telephone lines, which are more prone to errors, and the virtual protocol for the file can therefore also have errors. If the satellite office sends data in packet form and the CRC is polynomic and large, burst errors will not go undetected to the telephone facilities.

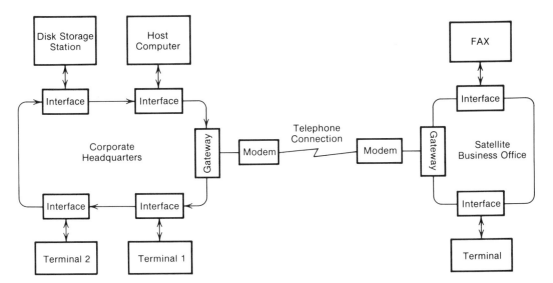

Figure 10–1 Simplified network architecture showing a remote bridge.

Virtual Terminals

Virtual terminals were briefly described in Chapter 8. The importance of virtual terminal protocols is reiterated here: they are protocols that hide incompatibilities in terminals from the layers below the presentation layer.

A simple example of a virtual terminal is as follows:

1. Text is prepared on a physical terminal; it is then translated by the presentation layer to a network virtual terminal form.
2. The data are then passed across the network to the destination.
3. The destination software then translates the data into the physical terminal local terminal format, which can be different from the source.

Some inherent problems associated with virtual terminals are that if a computer communicates with a terminal the task is fairly straightforward, but, for the other situation, the computer must have the ability to process data from all types of terminals connected to the network. The computer presentation layer will usually also have the task of performing the translations from terminal to terminal. With many low-cost microcomputers, particularly the 32-bit machines, much of the translation can be accomplished by the terminals.

Terminal-oriented protocols have several features in common. These commonalities must be incorporated as part of the terminal-handling process; they will be investigated in the following paragraphs.

Most terminals must have attention or interrupt capability. This type of function is performed out of band (i.e., outside normal data flow). These transmissions are not affected by flow control; they also have a priority status. Provisions are usually made so that, when the signal is processed at the destination, the meaning is clear to the program the terminal was communicating with.

Control words, such as carriage return, line feeds, rubouts, and backspace, that affect local terminal handling must be considered. For dial-up terminals, these parameters are not always known at network access point. The protocol must make provisions to set them. When terminals are permanently connected, the problems are somewhat reduced.

Dialogue between terminals may be different; that is, one terminal may operate half-duplex, another may transmit a line-at-a-time half-duplex, and still another may operate full-duplex with transmissions of a page of text at a time. All these techniques must be considered, and provisions must be made so that terminals with different dialogues can communicate. However, the number of different dialogues may be reduced because many installations use a limited variety of terminals. But if one wishes to design a truly universal presentation layer protocol, all possibilities must be considered.

The terminal data structure varies with the complexity of the terminal (e.g., scroll class as opposed to a fully microprocessor controlled data-entry type). As one can imagine, the task of making these two terminals communicate correctly is not trivial. The sophisticated CRT terminal (microprocessor controlled) has many commands, such as cursor addressing, line delete, and other edit features that the scroll terminals do not have. Not only may commands be different, but character coding may also be different; therefore, a break command on the two machines may decode as something entirely different when transmitted to a destination machine with coding differences.

The terminals must be capable of interacting with the application layers, terminal-to-terminal, process-to-process, and host-to-terminal. This implies that the protocol must have a symmetrical nature. One technique to observe true symmetry is for host 1 to appear as a terminal to host 2, or vice versa. Each host may use the program of the other machine through this virtual terminal technique.

When the terminals connect, they must settle their facilities' differences through negotiations. These negotiations must be dynamic, and issues such as parameter values must be determined. For example, all the commands, dialogue, and other presentation layer issues must be settled before data transfers can begin.

The last terminal characteristic to be considered is expansion capability (i.e., the protocol must be expandable). As new, more sophisticated terminals enter the market, which may be specialized (such as airline reservations and banking), the protocol must be capable of accommodating them. Also, the protocol must be backward compatible. An example of backward compatibility is when the protocol accommodates not only the new terminals but also the older machines that it was originally designed for.

Another approach to making incompatible terminals compatible is parameterizing their differences. Advocates of this approach are the CCITT (Europe, X.28, and X.29) and the Telenet Corporation (W.S.). This technique sets the parameters according to their function, whereas the virtual terminal approach is organized toward sets of parameters. For a more in-depth treatment of the parametric approach, see the X.28 and X.29 recommendations. The recommendations are not given in detail here because they are being upgraded and may be in some final form at the time this book is printed.

Let us now complete this study with a design example of a virtual terminal. The objective is to devise a virtual terminal supported by the presentation layer (i.e., the application layer handles its terminal as though it is a network virtual terminal).

The application program issues a command to the application layer, such as to move the cursor to the third line of text on the virtual terminal CRT. The sequence of events is that the application process updates its copy of the data structure, and the virtual terminal updates its data structure and display.

Some designs attempt to hide all the terminal idiosyncracies, but, as with language translation, all cannot be translated exactly or hidden. This is one reason the negotiations prior to any data transmission are very important. For example, most full-duplex terminals can operate in either full- or half-duplex. When negotiations occur between two terminals, the default condition shows half-duplex if one happens to operate only in half-duplex. If the parameter is somehow ignored during setup of the virtual terminal, the default condition will be favorable to reduce communications problems.

Let us examine some of the principles for designing virtual terminals. Option negotiations must be opened between the two virtual terminals to resolve issues such as screen size, character set, and class of terminal communicating. The negotiations can continue until all disagreements are settled or a determination is made that a connection cannot be completed. These connection failures are generally due to an applications program minimum that cannot be violated by one of the terminals.

The technique most virtual terminals use to establish a connection during negotiations is the following: Each virtual terminal transmits the desired criteria under which the connection is to be made. Then the two terminals set the initialization parameters to their minimum. If the connection cannot tolerate these minimums, it fails.

Break messages are a particular problem to be dealt with. If this command is allowed to be processed with normal transmission facilities, the reaction time before the printing stops on a standard printer can be time prohibitive, not to mention text continuing to be produced. As a case in point, a 30 character per second printer with a 2K queue can continue printing for approximately 1 minute after the break is invoked and the printed text will be one-half page. There are techniques of using interrupts to perform these functions out of band. The objective of any technique is to turn off the printer output and remove characters from the queue before they can be printed.

Let us examine a possible set of messages that can be used to implement a virtual terminal protocol. The message set is as follows:

OPCODE OPEN NEGOTIATIONS	MESSAGE LENGTH	NEGOTIABLE PARAMETERS

The message format shown is necessary to begin the virtual terminal dialogue. An opcode is necessary to determine the message type. This opcode may be a single byte or multiple bytes. The leading bit can be set or cleared when using alternate transmission modes (e.g., if either virtual terminal relinquishes its turn, the bit is set). Message lengths are self-explanatory. The field size may be one byte or multiple bytes depending on the number of parameters to be negotiated. The last field is the parameter to be negotiated, such as character set or full- or half-duplex operation. This message does not affect the lower layers of the ISO model because it is embedded in the data field of the process being addressed (virtual terminal). The message formats being described are a simple example of a protocol, but they may be further refined for an actual message set. The objective here is to give the reader some feel for the techniques used and the complexity of the message set.

The next message required must give the initial value of the virtual terminal parameters, the field size, and so on. A message format is as follows:

OPCODE DEFINE FIELD	MESSAGE LENGTH	FIELD SIZE	ADJUSTMENTS AND RENDITIONS OF PARAMETER	BOOLEAN PARAMETER	INITIAL VALUE OF THE TEXT

The first byte or character is the opcode for field definition, and again the first bit can be used for alternate operation (two opcodes required, one with the first bit cleared and the other with it set). The next field is again self-explanatory. The opcode defines the parameter and the field size will be in bytes. The fourth field describes the rendition of the parameter and adjustments to it. An extra field is included for Boolean expressions. The last field is where the initial values are represented prior to negotiations.

The reason for the previous two messages is to set up a virtual terminal for text transfer between two processes. A text message format is the next logical format, as follows:

OPCODE TEXT	MESSAGE LENGTH	DISPLAYABLE CHARACTERS

The first two fields are self-explanatory, and the last field contains some characters such as tabs, line feed, and carriage returns that may not be displayed. There may also be a set of characters that may require displaying only occasionally; they will require another message format, as follows:

OPCODE TEXT	MESSAGE LENGTH		OPTIONALLY DISPLAYED

This message may also include the on/off function as part of the opcode.

Control characters are also required. The message format is shown as follows:

OPCODE CONTROL	MESSAGE LENGTH	CONTROL CHARACTERS

The control message format is self-explanatory.

In the next message format, six single-byte messages are described, which are nothing more than opcodes.

Home   This opcode brings the cursor on the CRT to the first unprotected field at the top of the screen. The command is for cursor control, and again the first bit can determine which terminal is sending for an alternating transmission scheme. All other opcodes also require

	particular attention to the first bit for alternate transmissions unless specifically stated.
Clear	The clear command moves the cursor to the first unprotected field at the top of the screen and erases the contents.
NextU	This command moves the cursor to the next field; but if the U bit is set, it will go to the next unprotected field.
End of Transmission (EOT)	This command is sent when no more messages are to be sent; the acknowledgment must also be an EOT.
Interrupt	Self-explanatory.
Mark or Space	For filling in the field with blank characters.

The last four messages to consider are cursor moves, which are either absolute and require one or two fields, or relative, which can also require one or two fields. The four commands are as follows:

OPCODE ABS POS	POSITION PRESENT FIELD	NEW FIELD	Absolute Position

OPCODE REL POS	Δ OFFSET	Δ NEW FIELD	Relative Position

The opcode and the position move the cursor to an absolute position, and if new field is added (as shown in the dotted lines), the cursor moves to a new field at the position specified. For a relative move, the cursor is moved from its present position by the offset, which is a positive or negative number. This offset relative move can also be made in a new field by a field offset.

Fields as discussed here are protected or unprotected. Protected fields represent prompt title pages, comments for prompts, and the like, and unprotected fields may be where answers to prompts are entered or fields in which text is entered.

Opcodes may be single characters or a string of characters. Examples of the codes and their description are shown in Table 10–1.

The presentation layer must make the virtual terminal emulate ASCII displayable graphics if it is required to do so to meet the terms of negotiation. Often the ASCII graphics may be a part of a rendition, which may be a set of attributes such as reverse video, color for multicolor CRTs, or other similar character attributes. Several renditions may be available to select from, such as font sizes, print type, and color of characters. Some provisions may be made to adjust the rendition to better meet the virtual terminal requirements. For example, a character set that is a subset of ASCII may be required; this adjustment can be made to a rendition.

A typical communication between a virtual terminal and a computer (with virtual terminal) is shown in Figure 10–2. The terminal operator initiates the com-

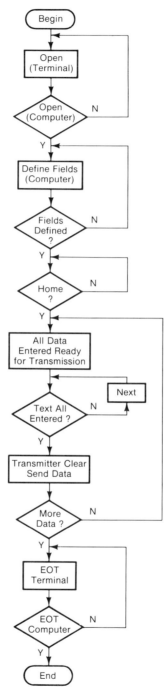

Figure 10-2 Flow chart of a virtual terminal communicating with a computer.

TABLE 10–1

Opcode (Hex) Representation	Description
81, 01	Open Negotiations
82, 02	Define Field
83, 03	Text
84, 04	Control
85, 05	Home
86, 06	Clear
87, 07	Next1
88, 08	Next0
89, 09	EOT
8A, 0A	Interrupt
8B, 0B	Mark
8C, 0C	Space
8D, 0D	Absolute Position
8E, 0E	Absolute Position New Field
8F, 0F	Relative Position
91, 11	Relative Position New Field

munication between the virtual terminal and the one embedded in the computer presentation layer. OPEN, in this situation, is opcode 81; the first bit of the opcode is set, indicating that the terminal is willing to relinquish its transmission (alternating mode half-duplex). The computer must send an OPEN (01); it does not relinquish its transmission, and the leading bit of the opcode is not set. The computer sends out a series of define field messages, which may be a form to be filled in by the operator. The computer will next send a HOME (05). The terminal process transmits all unprotected text (i.e., if forms are sent or prompts are sent, they are protected answers to queries, or data in forms are unprotected.) After the terminal has sent all the data alternating between TEXT and NEXT messages, the last message from the terminal relinquishes transmission; and the computer sends a CLEAR (86), which clears unprotected fields and relinquishes the transmission facilities to the terminal. The operator either begins entering data again or EOT (89), at which time the computer responds with an EOT (89, 09). The last response sent by the computer can be either EOT with or without the first bit set, because it is only used as an acknowledge.

Security Without Encryption

If a computer is located within a single room or installation, it can be made secure (i.e., no unauthorized persons can use it). Even in the military environment, this is often accomplished. The spaces where access can be gained to any of the computer equipment must be controlled with guards. Passwords are often used to prevent inadvertent access to files, but unfortunately many of them may incorporate a name or acronym that is commonly known. Uncommon words might be *#?,;!**, which is not to easy to remember. But passwords have been circumvented in many cases by computer hackers (the college or high school computer operator who enjoys breaking code words).

Another technique that holds a great deal of promise is voice recognition, which is in its infancy. But a combination of voice recognition and passwords or visual symbol recognition can be almost impossible to duplicate for intrusion purposes. The presentation layer could get complex to process all these functions.

Thus far the discussion has only addressed single enclosed computers or multiple computers located within a single enclosure with no connection to the outside world. However, LANs may connect computers, terminals, and storage stations located in several buildings and connected by cable. The weak link is the interconnect cable. The cable may be tapped, which breaches LAN security.

Fiber optics holds great promise for secure transmission because optical waveguides are very difficult to tap without detection. One reason for this is that severe geometrical changes will disturb the waveguide fields, and an optical time domain reflectometer (OTDR) can be used to locate the disturbance within 1 meter of where it occurs. Any switches, couplers, multiplexers, and the like, where crosstalk can be monitored, must be located inside secure controlled areas to prevent a breach of security. Alarm circuits can be installed in the fiber-optic transceivers with extra clad designed into the waveguide for alarm signals generated by optical field changes.

When national security is involved, elaborate measures must be taken. But when the information to be gained is less valuable than the security system protecting it, fiber optics may be a solution. The intrusion technique may only be known by a small group of people, which makes the probability of a tap small or nonexistent for most industrial applications.

Security with Encryption

Encryption is necessary when a LAN can be tapped or for national security. For example, if a LAN has a microwave, satellite, laser, or other link that broadcasts a signal, tapping is easily accomplished with a receiver. However, if the data are encrypted a key is necessary to decode them. With some techniques, the intruder does not even realize that data signals are present because they are buried in the channel noise.

Data encryption and decryption involve the technique of using a key (code word) and cipher (algorithm for transform action on a symbol-by-symbol basis) to encrypt the data. The data encryption standard (DES) forms the basis for cryptographic communications security for nongovernmental information. Encryption has been used successfully for Identification of Friend or Foe (IFF) for approaching aircraft.

To retrieve data that has been encrypted, a key is necessary to decipher the transmitted information. Deciphering is also performed by an algorithm on a symbol-by-symbol basis. The two keys (i.e., for encryption and decryption) must be identical to retrieve the data correctly. Keeping the key secret and known only to authorized users can become a monumental task when large numbers of users are involved. The probability of security penetration goes up with the number of users. For large numbers of users, the key becomes difficult to change because in all probability some of the users will not get the new key.

The key to decipher the encrypted data need not always be identical to the encipher key. It is possible, using transformation techniques, to make public one key for enciphering messages and to keep the other secret except for users with a "need-to-know."

Attacks on Encryption

Attacks on cryptosystems must be such that the intrusion goes undetected (e.g., a high impedance or inductive tap that defeats the physical system). If the intruder has unlimited computational power, the cost of an analytical attack on the system may be so great that it is not feasible. Also, the time required to decode the data may be too long for the intrusion to be of any value. If, for example, a military-type system is being dealt with, an air strike may be radioed to a pilot and the mission completed in 15 minutes. The intruder must decode the data within less than 5 minutes. On the other hand, during World War II, the United State broke the Japanese naval code and had days to decipher it prior to the attack on Midway. This is one instance of a successful attack on a code that produced significant results. In banking, where large amounts of money are handled every day, the possibility of security breaches is an immense problem.

The following discussion addresses a technique the cryptoanlyst may use for deciphering a text. The code may contain words that depend on the personality of the person or persons using the code. For example, the person devising the code word as a key may use his or her first or last name, children's names, the name of a girlfriend, and the like, which of course may be known by the intruder. The plain text (uncoded) may use one of several expressions to ask for the identification of the system user, such as PLEASE LOGIN, LOGIN, IDENTIFICATION PLEASE, or GIVE THE PASSWORD. One could possibly compile a list of 100 or so of these expressions; very few will use *?,;!"@ as a log-in expression. Also, a large number of words, phrases, ASCII, and other commonly used items are used frequently throughout the text. Examples are *the, and, or, etc., e.g.* Phrases such as *for example, such as,* and *therefore* are found throughout the text. ASCII characters such as spaces, line feeds, carriage returns, and punctuation are found throughout the text; for example, a standard 78 character/line text requires a line feed and carriage return that makes the plain text look rather cyclic. The encrypted text will not be cyclic, but should the intruder begin to note a cyclic nature to the text after trying a key, he will know he is getting close.

If the data encryption standard (DES) is used, which is a National Bureau of Standards hardware encryption technique, the key length is 56 bits and data are encrypted in 64-bit blocks. This algorithm (cipher) is produced in integrated-circuit form, which allows the intruder a little more time to experiment with breaking the code. The 56 bits allow $2^{56} - 1$ random key selections, but if one uses only ASCII characters, the key selections are narrowed. Many of these encryption techniques seem almost impossible to decipher, but today VAX 780 computer performance is available in microprocessor form. What may have taken one hour to run may soon be reduced to seconds or microseconds with the advent of molecular or atomic switching. This technique, discussed as early as 1966, for making switching elements with molecules or atoms would make the present microprocessors virtually obsolete.

The discussion presented in the preceding paragraphs gives the reader some insight into how a cryptoanalyst would begin to decipher an unknown code. The analyst may start by entering tables to be searched, such as books of names, street names, commonly used foreign words, or unusual words. At 2 million words or entries, at 80 or 90 words per minute (a slow typist's speed of entry), the time required to input the data would be 25,000 minutes or approximately 10 weeks at 40 hours per week. However, one must also remember, these tables may only need to be entered once and added onto occasionally. By using the DES technique, the analyst has two pieces of information to start with, the length of the key and that the data are encrypted in 64-bit blocks. The techniques for breaking this ciphering technique can be found in references [1] and [2]; they are beyond the scope of this book.

One precaution required by military encryption equipment is that it must be divided electrically into red and black compartments with strict guidelines as to the isolation. The red compartment is where data must pass before they are encrypted, which implies that the data must be kept in a secure environment. The black section is where the data pass after they have been encrypted, and security is not required here or at least it is minimal. Transmission of data then begins in the red section, is encrypted, and passes through the black section. The black (encrypted) data progress down the LAN to the receiver's black section and then are converted back to red (normal text).

When message traffic is to be sent encrypted, the destination address must be sent in the clear (i.e., plain text), broadcast encrypted, or one must use a different key from the main body of the text. When message traffic is sent in the clear, an intruder may measure traffic volume to a particular destination, which can be vital information. When the destination is broadcast in code, knowing which node on a network is receiving can assist the cryptoanalyst in attacking the cipher. If a separate key from the text is used for the destination address, even breaking the first cipher will only reveal the traffic volume the node receives, and the total encrypted message will not be compromised.

Stream ciphers can be made difficult to impossible to decipher by an intruder. Stream ciphers combine bit (byte) of the key on a bit (byte) basis with the data stream to encrypt the data. The key can be a random series of bits or bytes.

When the key is made as long as the longest message, the code is theoretically impossible to break.

Suppose the key is 64 Kbytes long; the key must then be carried somewhere because the operator can never remember a 64,000-digit word. Also, the time delay becomes large because all messages must be double their normal size.

Another method of generating a long key in a pseudorandom fashion is the use of a generator that is initialized with a small key. If the same key is used at the receive end, the same pseudorandom code is generated. One advantage of this type of encryption is that errors in bits during transmission do not hinder the receiver from deciphering subsequent transmission. Removal or insertion of bits in the data stream will affect the deciphering of these data. Stream ciphers do not lend themselves well to message-modification attacks. Autokey cipher can somewhat circumvent this problem; see reference [3] for more information.

Block ciphers similar to DES transform blocks of bits (64 for DES) under key control (56 bits for DES) into encrypted code (i.e., the clear text is mapped into ciphered text blocks). For a block size of n bits, the range of the clear-text values is 2^n with the number of $2^n!$ permutations. In actuality, all the permutations are not used because the cipher would become rather complex and the key size would be very large. For small block size, the weakness in the system is more apparent, because the plain-text block pattern can be compared with the ciphered. Therefore, the analyst may eventually find the key. Mixing techniques can be used on plain text before the data are encrypted. This will reduce the possibility of an analyst discovering the key. The CRC will provide a technique of producing what appears to be random code.

The block and stream ciphers provide the foundation for the bulk of encryption. Each has certain attributes that may be useful depending on the application. Block size in block cipher techniques is sometimes larger or smaller than the plain text. If plain text is larger than the block size, the text is simply broken into fragments that are the length of the block size. For the other situation (i.e., plain text smaller than block size), the block is padded with blank or mill characters. The extra characters waste bandwidth.

Block encryptions have identical patterns for identical plain-text blocks. Error propagation is also strictly intrablock, because these types of codes make the cryptoanalyst's work easier. One technique used to modify the cipher to prevent these two problems is called a chained block cipher. The encryption method is as follows:

1. The message is segmented into coding blocks and padded if necessary.
2. The first segment is enciphered.
3. This first enciphered block is combined with the second block of plain text using modulo 2 addition, similar to the CRC.
4. The result in step 3 is enciphered and the process is repeated.

Any errors that occur in the chained block cipher propagate throughout the block and all data which follow, and subsequent blocks remain unaffected. One objective of encryption is to produce codes that have substantial effects on cipher text for small errors; this results in difficult cryptoanalysis. Note that the only problem with the previous technique is that prefix blocks will match if the plain text is identical. Certain types of prefixes may be sent in the clear to prevent this problem, or the prefixes can be modified on a random basis to prevent this situation.

Autokey ciphers provide a key stream that can be matched to the text length that is to be enciphered; this eliminates plain-text padding. This, of course, improves bandwidth. Operating on 16- and 32-bit words will improve throughput as opposed to 8-bit words.

Often it is not enough to attack the code successfully but authentication is also necessary (i.e., the claimed identity of the principal is verified). Passwords, secret codes, voice recognition patterns or combinations of these can be used. If the authentication signature is known, the cryptoanalyst may have enough information to compromise security. When authentication signatures are transferred from one place to another, they must also be encrypted. In military applications, the signature (password) may be changed daily.

Often authentication codes have a long lifetime; for example, when using banking facilities, the customer's name is a lifetime signature. If some intruder could get this information and the necessary code, he or she could masquerade as the customer. Dynamic authentication is preferred as in the military (password-changing techniques). It is possible to do this with banking customers. For example, every ten times a customer at a bank uses his or her credit card for check cashing, the digital word of the card is modified, and this new word is sent back to update his or her authentication signature. This technique is rather costly, but the probability that the signature will be discovered by an intruder increases the longer the signature remains unchanged. For further information on cryptography, see reference [4], which has an in-depth study of cryptoanalysis techniques.

Data and Text Compression

One of the most immediate uses for text compression is for facsimile (FAX) transmission. For example, if a page is divided into 0.012 pixels (i.e., 0.012 × 0.012 squares), then an 8½ × 11 inch sheet will contain 650,000 pixels. If the pixels have 8 bits associated with their color, then the page would be represented by a 5.2-Mbit message. If the data were sent by switched network, the transmission time would be 9 minutes at 9600 bits/s. Group IV facsimile takes less than 1 minute to transmit a page of data. Therefore, text or data compression is necessary to achieve this transmission capability. If the message being sent has text and illustrations with a 1-inch margin around the document, compression techniques can compress margin, spaces between lines, large single-color spaces, and lines that appear in the direction of the scan. Also, if the document is black and white with no shading, such as a sales slip or order form, further compression can be performed.

There are two types of compression to be considered, symbol compression and facsimile compression.

Symbol compression can be accomplished by an examination of the symbol set to observe which are used more often. The most frequently used symbols should be the shortest in length. A table of symbols according to their probability of occurrence is shown in Table 10–2. These figures are for average usage; however, in a textbook on mathematics or for highly mathematical data where summations are used, i, k, z, and the like, may appear more often than normal. The table is approximate; adding up the percentages will not produce 100 percent.

TABLE 10–2
Probability of Use for the Alphabet

E	13%	H	5.4%	V	1%
T	10%	D	4%	K	0.5%
A	9%	L	3.8%	J, Q, X, Z	0.1%
O	8%	F, C	3%		
N	7.6%	M, U	2.5%		
R	7%	G, P, Y	2%		
I, S	6.4%	B, W	1.8%		

TABLE 10–3
Example of a Symbol Code

E	11111	S	11000	U	10001	K	00110
T	00011	H	10000	G	01000	J	10011
A	00111	D	11000	P	00010	Q	11001
O	01111	L	11100	Y	11101	X	10110
N	00001	F	11110	B	10111	Z	01101
R	11110	C	00100	W	11011		
I	11100	M	01110	V	01100		

For example, if the table symbols are encoded according to their use, as in Table 10–3, the symbols used the most have the fewest number of transitions from zero to one or from one to zero (i.e., T, A, O, N, R, I, S, H, D, L, and F have only one transition, C, M, U, G, P, Y, B, W, V, K, J, and Q have two, and X and Z have three). Let us now examine an ASCII code representation for "Now is the time for all good men to come to the aid of their country" in hexadecimal form (see Table 10–4). Table 10–5 is a comparison of ASCII with the new code for the expression "Now is the time." The number of transitions for the ASCII code is 61, while the new code has only 17. Single-bit transition such as occurs for ASCII S and R requires more bandwidth than the codes that have strings of zeros and ones between transitions. This can be calculated using fourier analysis on the waveforms. Also, for long strings of zeros and ones, a length code can be devised.

TABLE 10–4
ASCII Representation of "Now is the Time for All Good Men to Come to the Aid of Their Country"

	Now	is	the	time	for	all	
ASCII	4E,4F,57	49,53	54,48,45	54,49,40,45	46,4F,52	41,4C,4C	HEX
	good	men	to	come	to		
ASCII	47,4F,4F,44	40,45,4E	54,4F	43,4F,4C,45	54,4F		HEX
	the	aid	of	their	country		
ASCII	54,48,45	41,49,44	4F,46	54,48,45,49,52	43,4F,55,4E,54,52,59		HEX

TABLE 10–5
Comparison of Bite for "Now is the Time"

		NOW	IS
	ASCII	01001110010011101010111	0100100101010011
	NEW CODE	000010111111011	1110011000
		THE	TIME
	ASCII	010101000100100001000101	0101010001001001010000001000101
	NEW CODE	000111000011111	00011111000111011111

Text compression is another technique of compressing data; this can be accomplished using codes for common words in the language. As an example, 16 bits could be used to code 65,536 of the most common words in the English language; others can be encoded in ASCII. Still another possibility is to encode both words and phrases that are common and to use ASCII for uncommon words.

No space characters are inserted between words; they may be reinserted at the receive end. Also, the new code does not have punctuation or line feed and carriage returns. Methods of reducing the code field and expanding it again at the receive end are possible, which will reduce the total message size.

Conclusion

The primary aim of this text is to provide the reader with the knowledge of how fiber optics are incorporated into networks. Therefore, the majority of the book's emphasis is on the lower ISO model layers associated with the hardware. Other issues associated with this layer are time-sharing systems, data-base management, voice recognition, and visual pattern recognition. Textbooks have been written on each of these subjects. The last two subjects are in the research stage and are changing rather rapidly. The algorithms for the two types of recognition are continually being upgraded. One might expect voice recognition to be the most difficult, but a general statement cannot really be made because certain types of pattern recognitions are more difficult than voice recognition.

The presentation layer can also have directories, dictionaries, spelling correction, and other items as part of a library meant to assist the application layer and provide an interface to lower layers. The next and last layer in the ISO protocol hierarchy is the application layer.

REFERENCES

[1] M. E. Hellman, "DES Will Be Totally Insecure within Ten Years," *IEEE Spectrum*, vol. 16, July 1979, 32–39.

[2] M. E. Hellman, "A Cryptoanalytic Time–Memory Tradeoff," *IEEE Trans. Information Theory*, vol. IT-26, July 1980, 401–406.

[3] J. Savage, "Some Simple Self-Synchronizing Data Scramblers," *Bell Systems Tech. J.*, Feb. 1967, 449–487.

[4] A. G. Konheim, *Cryptography: A Primer*, John Wiley & Sons, Inc., New York, 1981.

11
APPLICATIONS LAYER

This is perhaps the most important layer to users because the boundary between application and presentation layer separates the user from the network designer. User programs can be quite diverse, which makes this particular layer difficult to describe because of large numbers of different users on the same network.

The discussions in this section will address the application-layer commonality issue among various users. When examining the application-layer representation to a LAN, the user will appear to be using a distributed computer. This computer can have a distributed operating system, data base, CPU mass storage, and computation capability. Each of these issues will be examined in more detail in the following paragraphs.

One of the more dynamic application-layer issues is array processing because of the continuing research in this area. Difficulties occur in array processing because program designers think in serial and parallel array processing, which is rather foreign to the normal thinking process. Some microprocessor manufacturers have produced components that emulate array processing, such as the Intel iapx 432 boards, which can be stacked to perform a limited amount of array processing. Other manufacturers are also producing multiple microprocessor systems. Array-type floating-point processors can be added to single or multiple processor systems with outstanding results; that is, the performance of these microcomputers can approach that of minicomputors, and minicomputers fitted with the array floating-point process approach the performance of mainframe computers. Therefore, networking is similar to array computers, with cables and fiber optics as the interconnections.

Distributed Data Base

The distributed data base can be a natural aspect of local-area network. For example, if host computers are used to store sales information locally and perhaps other hosts store manufacturing, shipping, maintenance, and data types, then a central computer using this information via the LAN will consider it a distributed data base. Also, each local host may desire additional information from the network hosts; this will become a distributed data base to the host desiring the information. Often a single host computer cannot handle the required amount of data; and multiple data storage is thus necessary. Disk storage stations can be used to relieve the host computers of some of these tasks. Storage stations are stand-alone nodes that usually employ a microcomputer to manage the subscriber traffic. These storage stations are ideal for situations where data are common to several hosts. The common data can be stored at these common sites. Each disk storage station can have multiple addresses, too.

Distributed data bases have three distinct advantages over centralized systems.

1. Reliability is improved because multiple computers provide backup service; therefore, catastrophic failure is rare.
2. Frequently used data may be stored in local storage, which promotes easier and quicker access.
3. Systems are modular and may be easily upgraded, which provides for future expansion of facilities.

The relational data base model will be discussed here; it is described in more detail in references [1] and [2]. Reference [3] provides the reader with some of the more controversial aspects of data-base management formats.

The relational data base model is usually defined as a rectangular matrix of stored information. The rows of this matrix (table) are tuples, with the attributes embedded in the tuple fields. The columns of the relation matrix (table) are called domains. Examples of these relational matrices for an airline are shown in Tables 11–1 through 11–5.

**TABLE 11–1
Flights**

Flight No.	*Origin*	*Destination*	*Type Aircraft*	
106	JFK	O'Hare	747	. . .
108	BOS	Heathrow	747	. . .
422	JFK	Shannon	727	. . .
627	O'HAR	Houston	737	. . .
411	BOS	Newark	DC10	. . .
. . .	. . .	. . .	. . .	. . .

TABLE 11–2
Reservations

Date	Flight	Booked	Name	...
7 April	106	210	John Smith	...
10 May	108	005	Tom Mones	...
11 May	422	115	Barbara Kurtz	...
1 June	106	311	Robert Lewis	...
12 Sept	411	412	Donald Baker	...
...	...	...	...	...

TABLE 11–3
Aircraft

Type	Registration	Seats	Flying Hours	Flight No.	...
747	11501H	450	350	106	...
737	11276B	100	210	108	...
DC10	10275H	270	970	422	...
727	10035P	85	227	627	...
747	20101J	450	512	411	...
...	...	...	...	...	...

An examination of the tables is in order to determine some of their characteristics. For Table 11–1, the domains are flight number, origin, destination, and type of aircraft. Each domain consists of attributes. For example, the series of flight numbers is referred to as attributes, with the field of the attributes labeled as a tuple (row). If the flight number is known, the other items, such as origin, destination, and plane type, can be derived. Therefore, this domain is considered a key. The keys for the other tables are Name (reservations relations), Registration (aircraft relations), Registry Number (maintenance relations), and Relation (dictionary relations). Each key, if known, allows access to the other domains' data. For example, in the aircraft relations, if the type of aircraft, seats, or flying hours domain is known, the particular aircraft cannot always be known, unless each is a different model. However, in large airlines a fleet of 100 aircraft of the 747 type may be in service, which, of course, makes identification impossible. Duplicates of tuples (rows) in this matrix are not permitted.

The dictionary is necessary to catalog the relations and provide information about each. For example, the flights relation is located in Chicago and has four domains and 1000 tuples. The tuple field size is 17 (i.e., 3 ASCII characters for three of the domains and 8 for the destination domain). This is not a very satis-

TABLE 11-4
Maintenance

Registry No.	Engines	Engine Hours				...
11501H	4	150	100	28	64	...
11276B	2	—	58	115	—	...
10275H	3	58	26	30	—	...
10035P	2	—	90	85	—	...
20101J	4	100	20	103	60	...
...	...		...			...

TABLE 11-5
Dictionary

Relation	Location	No. Tuples	Tuple Size	Domains	...
Flights	Chicago	±000	17	4	...
Reservations	London	300,000	30	4	...
Aircraft	Ireland	12	8	3	...
Maintenance	Boston	25	8	5	...
...	...	...	...	...	...

factory situation. It is desirable to have this information at each location. In the first case, the data were fully partitioned, and in the second case they are fully replicated. For an actual airline model, relations would be neither fully partitioned nor replicated, but some point between the two. For example, reservations data need not be stored in a maintenance area, or vice versa. A major problem with fully replicated data is updating; this requires all locations to be upgraded, as compared to fully partitioned data, where only a single update is required.

Data-base access allows queries about the information stored. Through queries, the subscriber can make updates to the data base. A Pascal-like high-level language is available (relational algebra). This technique describes how to find the answer to queries, (relational calculus describes the answer), but does not give the algorithm.

An example of relational algebra is a situation where two relations may be joined because they have a common domain, such as Tables 11–3 and 11–4. The resultant table is shown as Table 11–6. For this particular case, the key is the same (the aircraft registration number).

Two relations can be formed from one using projections. The single domain is an input, and the output is two specified domains as illustrated using Table 11–

TABLE 11-6
Aircraft and Maintenance

Type	Registration	Seats	Flying Hours	Engines	Engine Hours				Flight No.	...
747	11501H	450	350	4	150	100	28	64	106	...
737	11276B	100	210	2	—	58	115	—	108	...
DC10	10275H	270	970	3	58	26	30	—	422	...
727	10035P	85	227	2	—	90	85	—	627	...
747	20101J	450	512	4	100	20	103	6	411	...
...	...	...	...	...	...				...	...

TABLE 11-7
Aircraft

Type	Seats	...
747	450	...
737	100	...
DC10	270	...
727	85	...
...	...	...

TABLE 11-8
Aircraft (Characteristics)

Type	Registration	Seats	Flying Hours	Flight No.	...
747	11501H	450	350	106	...
737	11276B	100	210	108	...
DC10	10275H	270	940	422	...
727	10035P	85	227	627	...
747	20101J	450	512	411	...
...	...	...	...	...	

3, with Tables 11-7 and 11-8 as a result. When this projection is completed, Table 11-8 has domains used by maintenance, while in Table 11-7 the domains may be useful to reservations. Through manipulations of the tables, it may be possible to reduce the data base without reducing the information content by eliminating redundant domains.

In Table 11-7 the number of tuples is reduced as a result of the projection.

The number of tuples could have also been reduced in Table 11-8 if a subset had been formed based on a restriction. For example, in Table 11-2, Reservations, if all the May flights were put in one relation, only two tuples would appear in that table. Thus restrictions make searching the data base simpler because the number of tuples is reduced. However, the number of relations searched increases. Eventually, each relation becomes one tuple if the reduction process is continued.

A mathematical treatment of joining and projection will be presented to provide the reader with a more formal approach to these techniques. The relations are searched in response to a query. For this example, the query is which aircraft is close to its major overhaul (over 900 hours) and what routes will this affect.

1. T_1 = reduction of aircraft flights for aircraft with 900 hours of flight time
2. T_2 = join T_1 and flights
3. T_3 = restriction of T_2 by flight hours > 900
4. Result = projection of T_3 onto flights

The query will be expressed in relational calculus:

Range of A is aircraft
Range of F is flight
Retrieve A flight hours where
A. flight no. − F. flight no.
F. origin = JFK and F. destination = Shannon

This is a rather simplified problem because only two tuples must be examined.

We have only briefly mentioned queries, but they must be processed. The strategy of query distribution is to measure the response time and total bandwidth required. The response time during interactive periods is what the terminal operator observes, which is paramount. The bandwidth required for processing the query is a more important parameter because it affects both cost of processing and throughput.

The simplest approach to query processing is to have the host, with the query, request all the relations necessary to perform the processing sent to it, as shown in Figure 11–1(a). This would make the host processing the query appear as a centralized host. This technique will usually perform the necessary tasks correctly; however, other means that are more bandwidth efficient should be considered. Large files often may need to be transferred, which, of course, requires large bandwidth.

An improved technique for processing queries is to move large files as little as possible and make joins to reduce the amount of data in a move. An example of this technique is as follows: The query host could command reservations to send its relations to personnel for tuple reduction, as shown in Figure 11–1(b) (e.g., selecting flights piloted by married women, which could be only a single

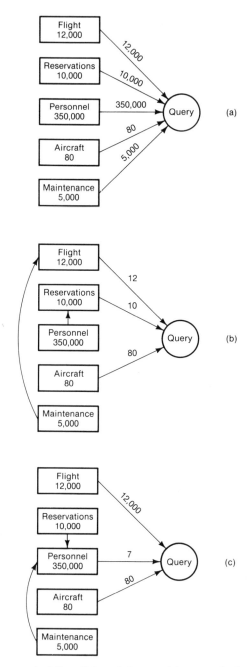

Figure 11-1 (a) Diagram depicting all the relations sent to query the host for processing. (b) Larger relations joined and the results sent to query the host. (c) Relations joined during the transfer of a large relation.

tuple). Maintenance and flight information can also be joined to find which aircraft are in check. The actual data to be moved to the query host are few compared to Figure 11–1(a). Note that the two joins occur almost simultaneously and reduce the processing required by the query host. If a large relation must be moved to the query host, then relations may be joined during the move to reduce some of the other large relations, as shown in Figure 11–1(c).

The data bases shown in the figures may be terminals with local storage, disk storage stations, or actual computers located in various departments. The connection may be local-area networks, telephone lines, microwave links, or satellite. Therefore, an analysis of the lower layers of the networks is necessary to determine the actual performance of the relation transfers. Long delays, for example, as exhibited by satellite communications, can have a large impact on whether to join two or more large relations. The technique to answer queries will be highly application dependent unless local-area networks are specified from the onset of the analysis.

For large local-area network application, concurrency control is necessary. As an example, if the same query is generated from multiple hosts, multiple updates can occur in the relations (reads and writes). The reader can well imagine the chaos and confusion that would occur. It is fortunate that there are techniques to circumvent this problem. One technique would be to allow only one transaction at a time, which eliminates concurrency errors, but system performance would be greatly impaired.

If a transaction is defined as a set of reads, then processing, and a set of writes, the formal approach described in reference [13] can be presented in discussing concurrency. In his approach, Bernstein defined a log as a line-ordered sequence of reads and writes performed on a data base as follows:

$$\text{Serial log}_i = R_i W_i R_{i+1} W_{i+1}$$
$$\log_{i+1} = R_{i+1} W_{i+1} R_{i+2} W_{i+2}$$
$$\log_{i+n} = R_{i+n} W_{i+n} R_{i+n+1} W_{i+n+1}$$

where $R = R(x_1, x_2, x_3 + \ldots + x_k)$ data items
$W = W(y_1, y_2, y_3, + \ldots + y_k)$ data items

Serial logs are transactions in which the write follows the read. As can be observed, serial logs form sets of logs. If reads and writes are interleaved, as follows, then the log is called interleaved.

$$\text{Interleaved log} = (R_1, R_2, R_3 \ldots R_n)(W_1, W_2, W_3 \ldots W_n)$$

Serialized logs can use a lock mechanism to lock out all other users when the data base is in use. This technique is used on MC68000 processors to secure the memory during read or write operations. It is a single assembly language instruction. A diagram of the technique is presented in Figure 11–2. Figure 11–2(a) is similar to a half-duplex connection where the data location can be operated

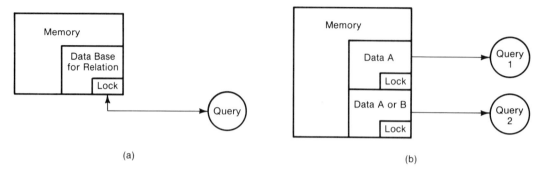

Figure 11-2 (a) Single memory location for a data base. (b) Dual memory location for a data base.

on by one user at a time. If the data base is used frequently or two relations are stored, Figure 11–2(b) can be the method of storage. In Figure 11–2(b), a duplicate storage may be used for data during heavy use periods, and if use decreases, the memory location may be reused for other data. This latter technique is, of course, rather complex and requires a more complex algorithm. The lock is nothing more than a flag that is examined and set if the data base is not in use. Each query must examine the flag before going into the data base. This technique makes no provision to prevent an application without the lock examination technique from accessing the data base while it is in use. Other methods with security measures may be implemented, but algorithm complexity will again increase.

Interleaved-type logs are prone to error, but serialized logs produce correct results. One objective is to serialize interleaved logs when possible to minimize errors.

There are a number of problems with data base locks that serialize data. Deadlocks, for example, are a problem. When query 1 has data required by query 2, and vice versa, each will attempt to lock the other's data base. The waiting period for such a transaction becomes infinite. One method is to allow a finite waiting period at which time both queries back off and await a random period before a retry. This is similar to CSMA procedures.

Token rings or virtual token rings are not prone to these problems because, when a process within a node is accessing another process within it, the LAN is not being used. However, the processes may set up a virtual token-passing scheme to prevent local deadlocks.

Only an overview of distributed data-base management is given here. Many new techniques will no doubt be discovered. The reader is urged to examine various trade journals for the latest developments.

Recovery from data-base failure (i.e., single-node failures) or the more catastrophic variety when the total system fails must be considered. Modern computers are generally highly reliable, and if failure does not occur quickly, such as a lightning bolt striking a machine, they will exhibit soft failures. A soft failure occurs when the machine has adequate time to store important data in nonvolatile memory before failure occurs. Recovery from a failure is as important as the failure itself, because both situations cause the system to undergo transient behavior. Examples of transient behavior are as follows: During the failure transactions could not be completed; therefore, nodes communicating with the failed node may keep trying in vain before the software discovers the failure. The second situation occurs when the node is reestablished and all the other nodes try to communicate to pass data. A danger of deadlocks will occur.

The failed node, when it recovers from a failure, can come up with a dummy transaction inserted in its processes that will prevent the node from being flooded with transaction requests. These dummy transaction requests can be removed slowly, which will allow the node to return to normal gradually. For the reader with a foundation in electronics, remember that systems disturbed by a step function often react in a damped oscillatory manner, as compared to systems disturbed by a comprising function with a shallow slope. The software of failed nodes responds in a similar fashion. For the reader who wishes to gain a more complete understanding of the procedures, see references [4–6] and *IEEE Transactions on Software Engineering*.

Array processors and accelerators are important topics that distribute and enhance computation performance of present-day computers. As an example, for large-scale computation, minicomputers can be fitted with a floating-point array processor and the computation performance will be close to a mainframe computer with a single user. These same techniques can be used on LANs with equally improved performance.

An example of distributed processing and computation is a situation involving a mainframe computer as a supervisory computer, with minicomputers provid-

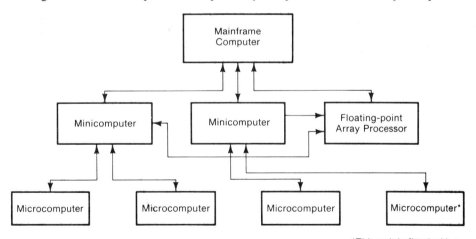

Figure 11-3 Computer hierarchical model.

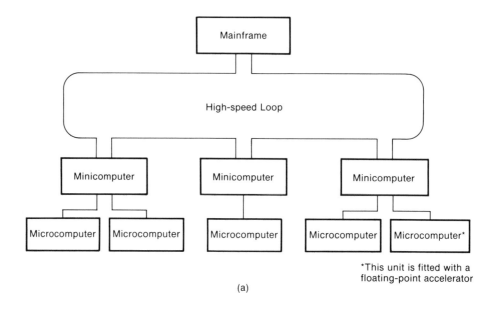

(a)

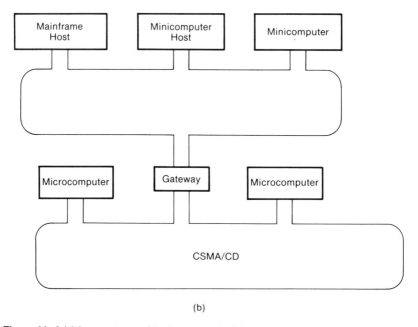

(b)

Figure 11-4 (a) Loop–star combination network. (b) Two-loop network.

ing multiple analysis capability, and a floating-point processor providing service for large-scale computation that would bog down the minicomputer or mainframe (see Figure 11–3). The figure depicts the minicomputers as receiving data to be processed from a series of microcomputers. To further increase the computing power of the system, the microcomputers that refine data to be further processed by the minicomputer may have floating-point accelerators added. This latter feature would be added only if the microcomputer has large numbers of computations to perform. The block diagram provides a simple overview. Let us now delve into a more complex model.

Before the complex model is discussed, a glance at the system will reveal that networking issues must be examined before the discussion can continue. For example, the mainframe may be attached to the minicomputer via a star LAN, or the minicomputer and the mainframe may be the center of a star network, with the microcomputers functioning as the satellite nodes. There are several possibilities. We must examine some of them and comment on the pros and cons.

The examples shown in Figure 11–4(a) and (b) have been discussed previously from a network point of view in Chapter 7. In Figure 11–4(a), the high-speed loop allows the mainframe to supervise the minicomputers that perform the computations for the microcomputers. The high-speed loop can be a 100MB/SEL fiber-optic-implemented network, which would allow fast transfers of large files. The microcomputers can be smart terminals or personal computers with preprocessing capability. When connected to a minicomputer as shown, these units form a star network with the loop connecting the stars.

The second connection, Figure 11–4(b), is a two-loop configuration with the high-speed loop having transmission rates of 100 Mbits/s. The second loop may be a low-speed loop depending on the required calculations being done by the minicomputers. The microcomputer that is put on a ring with the minicomputer will most likely experience reduced performance owing to congestion and the slower process time of the microcomputers.

Other configurations are also possible, for example, a single mainframe performing all the large calculations for the microcomputors, which are star connected to it. Another configuration might be multiple mainframes supervising multiple minicomputers, with the multiple mainframes acting as a backup for each other in the event of a failure. The possibilities are almost infinite. Therefore, a decision must be made about topology issues based on the application. The examples presented are rather simplified. A systems design approach should be performed to examine the performance issues, such as maximum acceptable delay for the transmission medium, maximum transmission speed, and throughput.

Often, minicomputers and microcomputers can be retrofitted with various printed circuit boards to upgrade their computational performance. In Figure 11–5(a), a typical mini/microcomputer is depicted with the various functional elements shown. The computer shown does not have a floating-point processor, which implies that the computation is done using a standard algorithm similar to

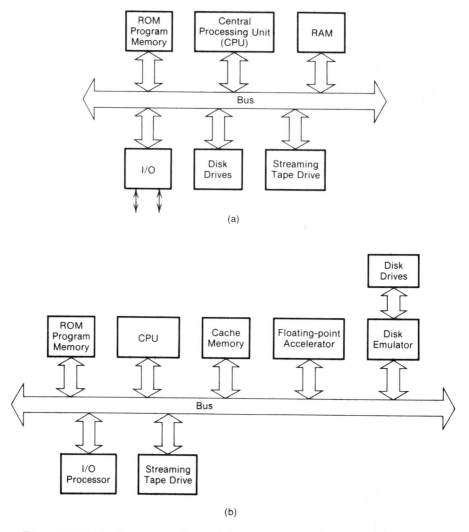

Figure 11-5 (a) Basic elements of a mini/microcomputer. (b) Modifications made to a mini/microcomputer that improve performance.

the one developed by the IEEE. All data are fetched directly from RAM, as shown. Also, disk data are fetched directly.

Adding the following board level modifications, which consist of hardware improvements, will improve the computational performance by a factor of from 4 to 6 (i.e., computation speed increases). The floating-point accelerator is a bit-slice hardware computational unit that performs arithmetic calculation approximately 10 times faster than the software version. The cache memory is a unit that prefetches data used in calculation from the RAM. The advantage of the cache is that access time to it may be 10 to 45 ns, as compared to the RAM, which can be as large as 800 ns for large storage space. A disk emulator (often called a RAM disk) prefetches data from the disk, which reduces disk access time by a magnitude or more. The final improvement is an I/O processor; this board performs the necessary I/O processing, which normally would interrupt the main CPU. Servicing the I/O interrupts increases the computation time of the CPU.

A question that will normally be asked about the modified computer is the following: Why is the improvement only four- to sixfold? In the discussion, mention was not made of the bus system. Bus process time may cause the computer to become bus bound, which implies that no matter how many modifications are added the bus will limit data transfer. So before adding modifications, examine the bus structure; this will prevent the designer from approaching the point of diminishing returns (i.e., large costs for hardware improvements with small increases in performance).

Software overhead can also increase, which will hamper projected increases in performance. For example, the main CPU may require a series of call instructions to access the floating-point accelerator. The call instructions may add sufficient overhead to reduce the performance increases of the accelerator. Some software techniques keep the calculations close together in the program, which can reduce the calculation time of the accelerator. The reader should be aware that careful programming can improve calculation performance. Also, when possible one should avoid series evaluation calculations, such as when evaluating exponentials. For example, evaluating X^3 requires longer computation time than the calculation $X \cdot X \cdot X$. The author noted a twofold increase in computation time of a benchmark program run with both X^3 and $X \cdot X \cdot X$.

Bus-bound systems are difficult to circumvent because they approach the CPU minimum access time. Some micro/minicomputers are equipped with dual bus systems. A local bus provides high-speed data transfer between the internal components of the computer, while a second bus provides data transfer between I/O, disk drive DMA (direct memory access), channels, and so on. The buses are connected through a bus processing unit. These systems will not be addressed in this book due to lack of space. Often one will even find multiple processors in implements such as terminals. A special CPU will usually be dedicated to keyboards, to monitoring video processing modem control, and to many other functions.

The discussion has addressed distributed computation in a rather simple and informal manner. As one may surmise, a topographical technique can be used to study distributed computation models. A method derived by Stone and Bokhari is given in the reference [7]. The method used to analyze problems in distributed computations has assumed a job consisting of a collection of modules using computers as nodes, which represent data sinks or sources, and the modules act as nodes. Load factors can be used to represent real running time on the branches between nodes. Then a graph of the network can be constructed. The analysis can be performed using any of the standard graph analysis techniques discussed in Chapters 2 and 7. The problems can be reduced to standard networklike problems.

Intuitively, the computers will see the tasks they perform as transparent to the networking. Therefore, we can assume that the behavior will be similar to a network consisting of intelligent nodes, which, of course, resembles star, ring, or bus networks, or a combination of the three networks.

For the reader who wishes to delve into other distributed computing systems, see reference [8]. A brief description of each technique is given in the following paragraphs.

The user–server model is a system of personal computers equipped with local storage. These personal computers may be equipped with several megabytes of memory and the local storage may be a 10- to 20-megabyte disk drive. If a system of these computers is devised with high-speed communication between them, this will allow each user to share data. For increased performance, several disk storage stations and perhaps a host and floating-point processor can be added, which can be shared by all the subscribers to the network. One advantage of these systems is that future expansion is rather easy to accomplish if provisions are made during the design. The provisions are sufficient address field size or a field with expansion capability, channel capacity large enough to accommodate the extra transmission loads, and the use of standard interface modules. Note that the issues are hardware rather than software, but as one may soon discover the two must be considered to have a viable system. Therefore, software issues must also be studied for compatibility.

The pool processor model described in reference [8] has terminals connected to concentrators, which is more efficient because terminals have rather low transmission rates compared to the computer network, an approximately 10 to 1 difference. The concentrators time share the same connection. These units are similar to the terminal servers of the Ethernet network. Several other pros and cons can be found in reference [8].

The third model used in the reference is a data-flow model. This technique uses a LAN as an array of processors that cooperate to run a single program. This technique is rather foreign to the average programmer because it does not follow the von Neumann machine model as constructed by John von Neumann (1946). Most people think serially (i.e., they look for a series of logical steps to perform tasks). Array (parallel) processing allows many steps to be performed at one time to achieve a result. Array processing is in its infancy and readers that are interested in array techniques should spend much time with the literature before contemplating a design. The topic will only be briefly addressed at the end of this chapter to stimulate interested readers.

Homogeneous Operating Systems

Operating systems are rather diverse, and the command language and words differ for various vendors. If all these various machine operating systems were connected together into one huge system, imagine the chaos that would occur—a virtual programmer's nightmare. To circumvent the problem, a network operating system can be devised that allows each machine to operate under its old operating system, but the network operating system is implemented as though the collection of user programs were running on various hosts. This technique is commonly called a network operating system. Another technique is to devise a distributed operating system to be used by all network implements. The first method is the simpler to implement, whereas the second requires less program and operating system maintenance. From previous experience, the author has worked with the former and operating system maintenance was an ongoing task because of glitches discovered under certain machine conditions. Finding one error and correcting it with a software patch resulted in two errors occasionally.

The first technique will be covered briefly because it is the least desirable of the two. The objective is to construct an operating system layer that may or may not hide all the operating system's differences. As one will note, this is consistently done through the ISO model to provide adequate interfaces between certain layers and to hide differences. The objective here is to hide differences in operating systems. One may ask the logical question: Why not use all UNIX machines? One reason is that many manufacturers have UNIX-like machines. Often implements are made with a manufacturer's internal standard, and the company may be large enough to give the potential customer a take it-or-leave it choice.

The network operating system task is to superimpose a process on the hosts that will provide a uniform interface to each host. The interface can be called INTRFAC. The process INTRFAC may run on the host's or a network access machine. If the process INTRFAC does not completely hide the existence of multiple and different hosts from the user, then commands may appear rather different, such as AT MACHX DO COMMUDY. The user must select the host each command is to be run on. The INTRFAC process must maintain a data base with sufficient information about each host and user's programs and data to allow user programs to run error free on the host chosen. INTRFAC is a command processor or process that translates user commands to the command language of the appropriate host. This type of system is fairly simple and does not affect the host computers. The hosts are unaware of the existence of the network or other hosts. The data files to be used by a particular host must be assembled prior to program execution. Also, interactive I/O operations are rather difficult because the program being run by the host must be informed in advance.

ARPANET has a project that is operated in this manner. It is called National Software Works (NSW). This technique allows a naive user to log on the NSW system and use programs and files distributed throughout ARPANET without being aware of their locations. Only NSW support programs can run that require compilers, assemblers, loaders, simulators, debuggers, and text editors; these are commonly called tools. Some examples of these systems can be examined in reference [8].

Distributed operating systems are the other alternative. This task is rather formidable compared to the previously discussed network operating system. Either of two models are often implemented by researchers.

In the process model, the various resources such as files, disk, printers, and other peripherals are each managed by some process, with the operating system that manages the communication placed between these processes. The second model is the object model. Each object is represented by a series of operations that it can perform. An operation may be performed on an object if the user process possesses the capability. The operating system task is to manage capabilities and allow operations on the object to be carried out.

When centralized operating systems are implemented, the operating system may be called on to provide security from malicious intruders. In these centralized systems, the capabilities must be distributed throughout the system and protected from intruder fabrication. Therefore, the capabilities may require encryption.

The two models are rather involved. It is suggested that the reader desiring more information on the subject consult the references [9–12].

In the network and distributed operating system, the assumption is that the von Neumann machine was used (serial processing); but as we will discover in the next section, parallel (or array) processing presents the operating system designer with a new multitude of problems. The subject will not be addressed here due to lack of sufficient information, but the reader should be aware of the next generation of these systems.

Parallel Processing

The reason this subject is addressed here is that parallel processing will have a large impact on future LAN design. As serial processing and serial LANs approach the point of diminishing returns (i.e., large cost increases for small improvements in performance), an alternative approach will be necessary to speed processing and transmission rates. Parallel (or array) processing and parallel transmission are the next logical step. Some examples of exploiting parallelism are as follows:

1. Performance of input, computation, and output operations on a given program simultaneously.
2. Initializing arrays to zero all at the same time.
3. Performance of vector or array arithmetic, for example, A ← B + C, where B and C are either vectors or matrices (APL uses this technique at present).
4. Simultaneous branching and bounding various nodes of a tree structure network or processing model.
5. Performance of the same sequence of operations on different sets of data at the same instant.
6. Performance of different and independent operations on the same set of data simultaneously.
7. Pipelining (or looking ahead), that is, simultaneous processing of several instructions in a sequential list of instructions, where they are in various stages of execution.

8. Performance of parallel search operation.
9. Parallel sorting.
10. Performance of simultaneous function evaluation, for example, in the search for a maximum or minimum of a function.
11. When transferring data from one node to another, transmission will be in words; therefore, large reductions in transmission time will occur.
12. Processing of data at the node will be in parallel; then nodal delays will be much smaller.
13. Queries must be made larger to accommodate the higher data rates, but megabyte RAM is in the offing at the time of this writing.

As one may observe from the list of advantages that some rethinking on architecture is needed. Some inroads have already been made to the parallel processing arena. The U.S. Department of Defense (DOD) has sponsored the development of a programming language called Ada. This programming language is useful for embedded computers (a system that is part of a larger more complex system). Embedded computer systems are large and have requirements for parallel processing, real-time control, and high reliability. Some present-day machines are produced with Ada operating systems. For example, the Intel iAPX 432 has an Ada operating system, but it requires a VAX 11/780 as a software development system. From past reports, it has been quite cumbersome, but this is only a start. The Hawk 32, a military-grade computer built by Norden, also has an Ada operating system. This particular computer is fully militarized (i.e., ruggedized and built into a militarized case). It also requires uploading and downloading from a large development computer. Eventually, Ada machines will be equipped with silicon embedded Ada.

Parallelism in computers is not without its trade-offs. The complexity of handling data in parallel will add overhead to processes and programs. This added overhead will reduce some of the throughput benefits of parallel programming. For example, the bit edges of a word or byte traveling along a parallel transmission path will not arrive simultaneously at the receiver end. Some compensation technique must assure that the bits are processed together. The CRC checking at the receive end is also more complex, but there are integrated circuits that will perform this task.

When a computer process is performed in parallel, the data and tasks are distributed to several computers that perform their assigned tasks, and after the tasks are completed the resultant data must be reassembled for presentation. The disassembly and reassembly of data require overhead, which again reduces throughput.

Parallel progamming language consideration will be examined to observe what is required to implement a parallel algorithm. The implication here is that a language to implement is available and the processors necessary to implement the algorithm are also present in the system.

As a starting point, a flow chart is given in Figure 11-6. The assumption is

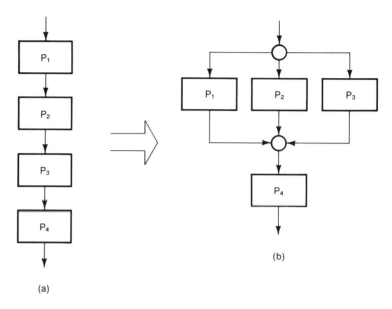

Figure 11-6 (a) Serial processor and (b) equivalent parallel processing.

that P_1, P_2, and P_3 can be executed simultaneously and P_4 can only be executed after the first three have finished. The flow chart indicates that P_1, P_2, and P_3 are independent. Each block will have typical statements or instructions such as assignments, conditional statements, DO loops, and subroutine calls. The exception will be jumps out of or into one of the parallel P_i's, which would be difficult to implement.

An example of parallel processing is given next. An IVTRAN (FORTRAN-like program language designed for the Illiac IV computer) statement is given as:

$$\text{DO } k \text{ CONC FOR ALL } (i_1, i_2, i_3 \ldots i_j) \epsilon S$$

where k is a label that defines the range of the DO loop, $(i_1, i_2, i_3 \ldots i_j)$ is a sequence of integer index variables, S is a finite set of n-tuples of integers that defines the range of index variables.

The intent here is to assign one processor to each j-tuple, for example $(v_1, v_2, v_3 \ldots v_j) \epsilon S$. The assigned processor will execute all the statements within the range of the DO-loop under the assumption that $i_1 = v_1$, $i_2 = v_2$, $i_3 = v_3$, $\ldots, i_j = v_j$. Simultaneously, other processors, corresponding to different j-tuples, will execute the same statements within the DO-loop.

Let us now consider an example of a code segment that computes the (SQRT) of each element of a 3 by 6 array B and places the result in A.

DO IO CONC FOR ALL (I,J)ϵ[1≤I≤3, 1≤J≤5]A(I,J) + SQRT(B(I,J)) IO CONTINUE

This concurrent DO-loop requires 18 different processors, each of which will execute the DO-loop in an asynchronous manner (i.e., the processors need not execute the same instructions simultaneously). This implies truly independent calculations.

Occasionally, synchronous operation is desired. This can be accomplished with the following expression:

$$\text{Do } k \text{ SIM FOR ALL } (i_1, i_2, i_3 \ldots i_y) \epsilon S$$

This expression is executed in parallel with SIM replacing CONC in the statement, which makes execution synchronous.

The Illiac IV computer will also excute instructions sequentially with the following DO-loop:

$$\text{DO } k \text{ SEQ FOR } K\epsilon[1 \leq K \leq 10]$$

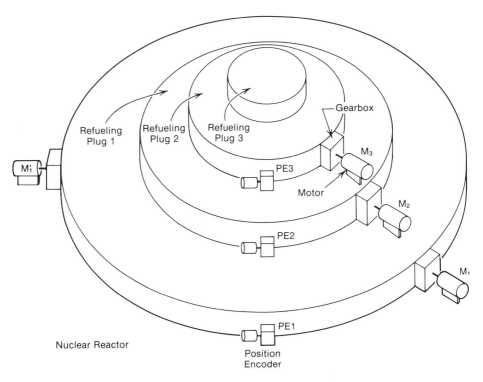

Figure 11–7 Nuclear reactor refueling system.

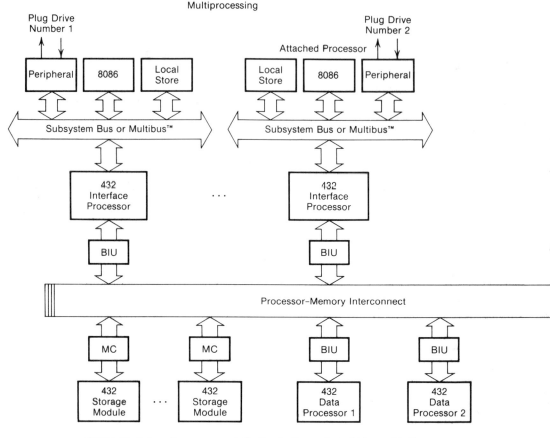

Figure 11-8 A multiprocessor application to the system of Figure 11-7.

A parallel series set of instructions can be expressed for multiplying two 10 by 10 matrices, A by B, which will require 100 processors:

```
              DO 200 CONC FOR ALL (I,J)ϵ[1≤I≤10, 1≤J≤10]
              DO 100 SEQ FOR Kϵ[1≤K≤10]
                  C(I,J) = 0
                  C(I,J) = C(I,J) + A(I,K) + B(K,J)
   100        CONTINUE
   200 CONTINUE
```

The program shown allows the reader some insight into what is involved in array processing, which is undergoing rapid change. Therefore, the reader desiring more information on the subject should search the IEEE and computer journals.

Applications Layer

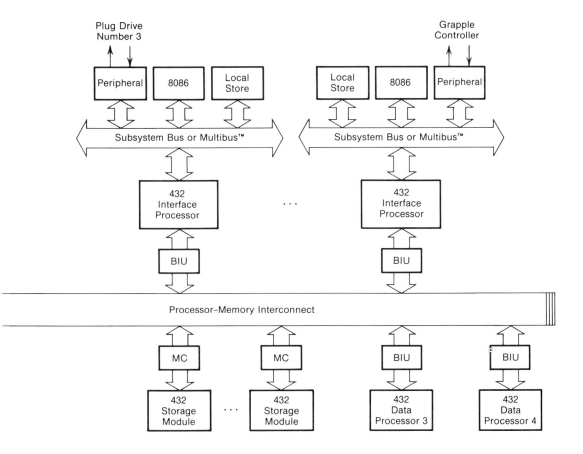

As a final topic, let us examine an overview of the Intel iAPX 432 hardware, which is an example of a microcomputer array processor. Often, for real-time processing of control function, this type of architecture is ideal. The controller for a reactor control system will be discussed. This is a generic application that can be applied to a common three-axis system with six degrees of freedom.

The reactor refueling system is shown in Figure 11–7. The objective of the iAPX 432 array processors is to control the plug drives M_1, M_1', M_2, and M_3. At first glance, the system appears to be a rather simple, straightforward control system, but there are some idiosyncracies that can cause trouble for even the best of engineers. (The wiring to the motors, position encoders, alarm switches, refueling grapple, and semic pins and also the grapple drawing, are not shown.) All the motors must accelerate at a constant acceleration to prevent excessive g loads on the structure (i.e., the velocity is nonlinear), which must be controlled by a computer. Each plug has the same criteria. The gears in the gearbox are very large and have a great deal of backlash. During start-up and stopping of the plug, the backlash must be taken out gently to prevent large torque spikes from occurring, which can cause damped oscillations in the system with disastrous results. Alarms must also be monitored to prevent the plugs from overdriving and twisting control wiring to the breaking point. Some interaction between plugs will occur during starting and stopping. The computers may be required to compensate for these interactions. Drive motors M_1 and M_1' must load share to prevent one motor from producing drag on the other. The position encoders are digital; therefore, position calculations must be made to acquire correct positioning. All the items mentioned are controlled in real time. One may logically ask: How is this related to a local-area network? The answer is simple: it is a very large system that must be interconnected with radiation-hardened cables. The control system is supervised by multiple hosts because a backup system is required for safety. The LAN may consist of components that meet either industrial or manufacturer's standards. The three motor control computers may be part of a parallel computer function, with the three position encoders as inputs and current monitors for torque. The objective here is to show an application for parallel processing. The author has been involved in a project similar to this, and a complete design would require several volumes to document. Therefore, only the parallel computer will be considered here.

Each of the motor control and positioning functions is part of the multiprocessing system shown in Figure 11–8. The peripheral input/output consists of motor current monitor inputs, sensor inputs position alarm, Geiger counters and the like, switch condition monitors for the control room, and control motor velocity outputs (analog). The local computer, an Intel 8086, functions as the motor controller with the local storage, a combination of E^2PROM and RAM. The E^2PROM stores the velocity curves and the control program. The 432 interface processor performs the necessary processing of status and control data to the data processor 432s to allow them to correctly supervise the system. Note that the system consists of four data processors that perform control function and also provide a backup in the event of single- or multiple-processor failure. The cabling to the system is either copper or fiber optic that is radiation hardened. This system allows the 8086 microprocessors to operate in parallel with the 432s supplying the necessary supervisory functions. The grapple controller is also supervised by the 432 processor. Therefore, three of the processors will operate in parallel and the fourth must operate after the three have finished their tasks or the grapple will be twisted off or otherwise damaged. Other items not shown can also be supervised and controlled in a similar manner, such as fuel handling cars, storage pools for spent fuel rods, and container handling areas. All these areas require remote operation of vehicles, robots, monitors, and alarms because of the radiation hazard. The control systems and networking are quite complex and require a very high degree of reliability. Part of their complexity is due to redundancy, and the remainder is due to the remote nature of all the functions that must be performed.

This final chapter of the book should have stimulated the reader's interest in application-layer subject matter because this is the layer most people are familiar with. The software person may want to begin with this chapter and work his or her way down to the physical layer. On the other hand, a person more familiar with hardware will follow the normal flow given in the book. The reader should now be equipped to read the many trade journals and periodicals with ease.

REFERENCES

[1] D. D. Chamberlin, "Relational Data-Base Management Systems," *Computing Surveys* 8(1), March 1976.

[2] E. F. Codd, "Further Normalization of the Database Relational Model," in *Database Systems,* Cowrant Computer Science Symposium 6 (R. Rustin, ed.), Prentice-Hall, Inc., Englewood Cliffs, N.J., 1972, p. 3364.

[3] S. B. Michaels, Jacob Millman, and C. R. Carlson, "A Comparison of Relational and CODASYL Approaches to Data-Base Management," *ACM Computing Survey* 8(1), March 1976.

[4] E. F. Codd, "A Relational Model of Data for Large Share Data Banks," *Commun. ACM,* vol. 13, 1970, 377–387.

[5] D. D. Clark and L. Svobodova, "Design of Distributed Systems Supporting Local Autonomy," *Compcon,* 1980, 438–444.

[6] Berkeley Workshop on Distributed Data Management and Computer Networks.

[7] H. S. Stone and S. H. Bokhari, "Control of Distributed Processes," *Computer,* vol. 11, July 1978, 97–106.

[8] A. S. Tannenbaum, *Computer Networks,* Prentice-Hall, Inc., Englewood Cliffs, N.J., 460–476.

[9] J. E. Donnelley, "Components of a Network Operating System," *Computer Networks,* vol. 3, Dec. 1979, 389–899.

[10] J. K. Ousterhout, D. A. Scelza, and P. S. Sindku, "Medusa: An Experiment in Distributed Operating Systems Structure," *Commun. ACM,* vol. 23, Feb. 1980, 92–105.

[11] R. W. Watson and J. G. Fletcher, "An Architecture for Support of Network Operating System Services," *Computer Networks,* vol. 4, Feb. 1980, 33–49.

[12] M. H. Solomon and R. A. Finkel, "The Roscoe Distributed Operating System," *Proc. Seventh Symp. on Operating System Principles ACM,* 1979, 108–111.

INDEX

A

Ada, 273
Adaptive routing, 166
Addressing, 161, 202
Analog phase-locked loop (PLL), 91
ANSI (American National Standards Institute), 17
APRANET, 150–151
ARPANET, 20–21, 223–227
Arpanet TCP protocol, 215

B

Backbone networks, 62–63
Bandwidth (BW), 76–78
Bessell functions, 75
Bit error rate, 101
Bit stuffing, 71
Broadband cable plants, 92–97
Broadband fiber-optic networks, 207–209
Broadband network, 205–207
Bus-bound systems, 269

Bus configuration, 33
Bus interface units (BIUs), 95

C

CCITT, 17
CLSN/CLSN, 91–92
CPU (Central Processing Unit), 269
CSMA (Carrier Sense Multiple Access), 79, 138, 172, 182
CSMA/CD (Collision Detect), 79, 138, 172, 182
Cable interface, 90–91
Cable plant design, 78–83
Channel capacity, 55–56
Copper cable, 112–114
Cyclic redundancy checking (CRC), 13, 114–115

D

Data encryption standard (DES), 247
Data-link controller (DLC), 149

Data transfer, 160
Data gram, 161–162
Decoders, 121–122
DECNET, 24, 152–153, 228
DECNET Router Server, 223
DELNI, 184
Digital data communications message protocol (DDCMP), 152
Digital Equipment Corporation (DEC), 5, 24, 184
Digital-to-analog converter (D/A), 96
Direct memory access (DMA), 200
Directory routing, 163
Distributed data base, 257–270
Distributed processing, 265
Domains, 257
Dynamic routing, 163

E

E^2PROMs, 47
ECMA (European Computer Manufacturing Association), 17
EIA (Electronic Industry Association), 17
EIA standards, 80
EXOR, 121
Electromechanical positioning system, 179
Electronic mail, 6
Encryption, 246–251
Erasable read-only memories (see E^2PROMs)
Error control, 161
Error detection, 217–220
Ethernet, 24–26, 80, 84–90, 153–154
Ethernet serial interface (ESI), 89

F

Fiber optic cable plant, 97–105
Fiber optic transmitters and receivers, 106–112
Fiber optics, 112–114, 207–209

File transfer, 238–239
Flooding, 163
Flow control, 161, 202, 218
Frame check sequence (FCS), 115–117
Frequency-division multiplexing (FDM), 10–11
Frequency modulation (FM), 72
Frequency shift keying (FSK), 72
Full-duplex, 160–161

G

Gateways, 201–205

H

High-level data-link control (HDLC), 148–150
Host computer, 182
Hybrid star network, 197–200

I

IBM, 5
IEEE (Institute of Electrical & Electronic Engineers), 17
IEEE 802 Local Network Standards Committee, 143
I/O processor, 269
Identification of Friend or Foe (IFF), 247
Intel Corporation, 89
Intel ESI B2501 IC, 92
Interface, 161, 271
Internal protocol (IP), 25
International Business Machines (see IBM)
International Consultative committee on Telegraph and Telephone (see CCITT)

Index

285

International Standards Organization (ISO):
   comparison models, 21
   model of networks, 6, 10–18
   standards, 17

L

LAN Information Theory, 67–78
Least significant bit (LSB), 133
Light-emitting diode (LED), 97
Local area networks, selection of, 211–212
Local-area transport (LAT) protocol, 222
LocalNet, 27
LocNet 20, 95
Loopback, 92
Loss budget checkoff list, 99–100

M

Manchester coding, 120–121
Manchester encode data, 90
Metal oxide varistors (MOVs), 80
Microprocessors, 191, 198
Modems, 77, 96–97
Monte Carlo Method, 47
Most significant bit (MSB), 134
Motorola Mc68000, 28
Multiplexers, 122–126

N

NBS (National Bureau of Standards), 17
NCP (Network Control Protocol), 223–227
National Software Works (NSW), 271
Network Control Center (NCC), 28
Network transparency, 160
Networks:
   flow of information, 38–48
   two-loop, 3, 5
Node, 37

Noise, 56–61
Noise filter, 90–91

O

Operating systems, 270

P

Packet communication unit (PCU), 235
Packet radio, 209–211
Packet switching, 20–21, 228
Parallel processing, 272–279
Pascal, 39, 41–43, 46–47
Password (*see* Encryption)
PIN description, 93
Point-to-point wiring, 3–4
Poisson's distribution, 53
Port technique, 216
Protocols, 25, 131–136, 155–157, 238–239

Q

Query processing, 261

R

R (local repeater), 182
R (Remote repeater), 183
Real-time applications, 219
Ring structures, 172–182
Rise time calculations, 104–105
Router server, 184
Routing, 163–171

S

SNA (Systems Network Architecture), 22–22–24, 231–234
SNA gateway, 184
Satellite communications, 209–211
Security, 220, 246–251
Sequencing, 162
Serial-to-parallel conversion, 197
Set next node, 189
Signaling, 161
Sockets, 215
Specifications, 155–157
Star networks, 191–197
Switching, 180–181
Symbolic compression, 252
Synchronization, 219
Synchronous data-link control (SDLC), 148–150
Sytek Inc., 27, 77, 235–236

T

TCP (Transmission control protocol), 25, 223–227
Teleconferencing, 220
Terminal server, 184
Text compression, 251–253
Time-delay, 48–54
Time-division multiplexing (TDM), 10–11
Time-division statistical multiplexer (TDSM), 122–126

Token bus, 189–191
Token passing, 185–188
Token rings, 188–189
Topology, 33–36
Transmitter encoding, 120–121

U

USRTs (Universal synchronous receiver transmitters), 197
UNIX operating system, 25

V

Verifications, 155–157
Virtual terminals, 222–223, 240–246

W

Wang, 5
Wagnet, 26–29
Western Digital, 71

X

X.25 standard, 20, 29, 228
X.25 gateway, 184
Xerox, 24–26